GEOMORPHOLOGY
AND
HYDROGEOLOGY

A Handbook

GEOMORPHOLOGY
AND
HYDROGEOLOGY

A Handbook

S.M. HAMID RIZVI
MSc, DHMSC, MPhil

Practising Geologist

CBSPD

CBS Publishers & Distributors Pvt Ltd

New Delhi • Bengaluru • Chennai • Kochi • Kolkata • Lucknow • Mumbai
Hyderabad • Jharkhand • Nagpur • Patna • Pune • Uttarakhand

Geomorphology and Hydrogeology

ISBN: 978-81-239-1547-0

Copyright © 2008, Author & Publisher

First Edition: 2008

Reprint: 2009, 2015, 2024

Published by Satish Kumar Jain and produced by Varun Jain for

CBS Publishers & Distributors Pvt Ltd

4819/XI Prahlad Street, 24 Ansari Road, Daryaganj, New Delhi 110 002, India.
Ph: 23289259, 23266861   Website: www.cbspd.com
e-mail: delhi@cbspd.com

Corporate Office: 204 FIE, Industrial Area, Patparganj, Delhi 110 092

Ph: 4934 4934   Fax: 4934 4935   e-mail: publishing@cbspd.com; publicity@cbspd.com

Branches

- **Bengaluru:** Seema House 2975, 17th Cross, K.R. Road,
  Banasankari 2nd Stage, Bengaluru 560 070, Karnataka
  Ph: +91-80-26771678/79   Fax: +91-80-26771680   e-mail: bangalore@cbspd.com
- **Chennai:** 7, Subbaraya Street, Shenoy Nagar, Chennai 600 030, Tamil Nadu
  Ph: +91-44-26680620, 26681266   Fax: +91-44-42032115   e-mail: chennai@cbspd.com
- **Kochi:** 42/1325, 1326, Power House Road, Opp KSEB, Ernakulam 682 018,
  Kochi, Kerala, India
  Ph: +91-484-4059061-65   Fax: +91-484-4059065   e-mail: kochi@cbspd.com
- **Kolkata:** 147, Hind Ceramics Compound, 1st Floor, Nilgunj Road, Belghoria,
  Kolkata 700 056, West Bengal, India
  Ph: +91-33-25633055/56   e-mail: kolkata@cbspd.com
- **Lucknow:** Basement, Khushnuma Complex, 7-Meerabai Marg
  (Behind Jawahar Bhawan), Lucknow 226 001, UP, India
  Ph: +0552-4000032   e-mail:tiwari.lucknowi@cbspd.com
- **Mumbai:** PWD Shed. Gala no. 25/26, Ramchandra Bhatt Marg, Next to JJ Hospital
  Gate no. 2, Opp. Union Bank of India, Noorbaug, Mumbai 400 009, Maharashtra, India
  Ph: 022-66661880/89   e-mail: mumbai@cbspd.com

Representatives

- **Hyderabad** 0-9885175004 • **Jharkhand** 0-9811541605 • **Nagpur** 0-8692091830
- **Patna** 0-9334159340 • **Pune** 0-9664372571 • **Uttarakhand** 0-9716462459

Printed at SRK Graphics, Delhi, India

Dedicated to

---

*the sweet memory of my
beloved father*

Shaheed S. M. Mustafa Rizvi

*who was a
great social worker*

Preface

Earth is full of miracles but many myths can be explored after studying the subject of geology. Geology as a subject of study is a must for all those who want to know the past, present and future happenings occurring *on, above* and *underneath* the Earth.

I take this opportunity to present to you

🐾 | GEOMORPHOLOGY AND HYDROGEOLOGY : A HANDBOOK |

which gives the facts and necessary information in such a way so to serve as a very useful *text* for the students of geology, and as a reference *handbook* for the practising geologists, engineers and professionals concerned with this and other related fields.

The *Handbook* has been written to cater to the needs of geology students of university classes. I have tried to present a practical approach of drilling, which is one of the most bright future areas for the geology students. My working tour to Sultanate of Oman was to get practical experience in the field of drilling and its environmental approach that I have included in this book very succinctly. This book will prove highly useful to the students of BSc (Hons) and MSc in geology and also useful for further study and research in the field of geology.

Some objective style questions on geomorphology and hydrogeology have also been given at the end of the text of each section.

The whole credit of writing this book goes to my father and mother (late Mrs Zakia Khatoon) whose inspiration and encouragement were solely responsible for my throughout good academic carrier. I feel immense pleasure in taking the opportunity of acknowledging my sincere thanks to my wife Mrs Urooj Fatma Rizvi, Mr. S.M. Sarim Rizvi (son), Md. Syed Shakil Ahmad (maternal uncle), and Mr. Syed Khalid Saifullah and Mr. S.M. Majid Rizvi (brothers) for their encouragement in writing this book.

I wish to express my deep gratitude to Dr. Shadab Khurshid, Reader, Department of Geology, Aligarh Muslim University, Aligarh. I am also grateful to my educational guru Hon'ble Justice Syed Aftab Alam, Judge at the Patna High Court.

My special thanks to Mr. S.K. Jain, Managing Director, and Mr Y.N. Arjuna, Publishing Director, CBS Publishers & Distributors, for their active cooperation in bringing out this book in a presentable form.

S. M. HAMID RIZVI

Contents

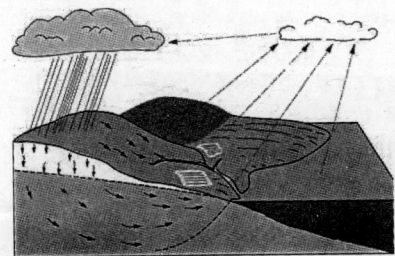
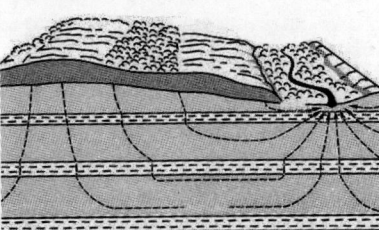

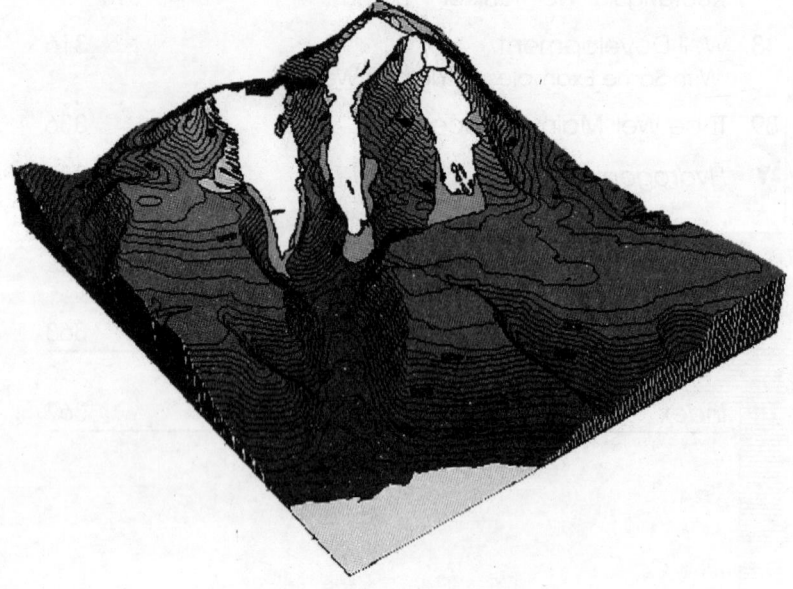

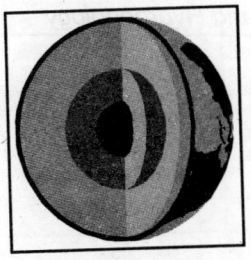

Part I

GEOMORPHOLOGY

GEOMORPHOLOGY

Geomorphology is the branch of geology that examines the formation and structure of the features of the surface of the Earth. It is the study of landforms, including their origin and evolution, and the processes that shape them. It seek to understand landform hostory and dynamics, and also predict future changes through field observation, experimenting and numerical modeling. The landscape is built through tectonic changes and volcanism; is also lowered by subsidence due to tectonics or physical changes in underlying sedimentary deposits.

Practical applications of geomorphology include measuring the effects of climate change, hazard assessments including landslide prediction and mitigation, river control and restoration, coastal protection, and assessing the presence of water.

Part I

GEOMORPHOLOGY

1 Beginning and Scope of Geology

Geology is purely a science subject. The word *geology* was coined by a Bishop in the fifteenth century. Geology, *geo* means earth, *logos* means science, which deals with the earth as a whole. Geology reveals the hidden treasures of the earth.

It is interesting to mention here that early man was curious to know, what was the mechanism involved in the natural happenings on the earth and why earthquake and volcanic eruptions took place.

In fact, in early times the study and practice of geology received serious setback due to the religious faiths and beliefs followed by Church in Europe and by the Muslims in Asia.

Earlier, great men like Omar Khayyam and Leonardo da Vinci found some fossils on the basis of which they opined that these fossils were the remains of living (marine life) which remained buried since the regions had been beneath the sea, but their ideas about the earth were condemned by the other people of the time.

In 1776, the year of Independence of America, Werner, a scientist from Germany, discovered fossils and said that the different layers of the rocks could be identified by the fossils that occurred at a particular horizon.

It is important to note that geology in its early history was only as an academic science, but the industrial revolution brought a turning point in the subject as all types of raw materials were required in abundance for various types of industries. Geology and the geologist then started contributing to the treasures of mankind.

Science of geology is a vast subject. Generally, three broad subdivisions of geology are seen, viz. **general geology**, **petrology**

and engineering geology. In this book only two branches of geology are discussed, which are **geomorphology** which describes the major relief features of the earth and **hydrogeology** which mainly covers the description of environmental control of water with special features as drilling. It also includes the study of atmospheric water, surface water and underground water, which are the subjects of national and international interest. Geology begins with the observation of global natural changing characters with unlimited scope in the field.

The Scope of Geomorphology

Geomorphology is the systematic study of landscape. A few scientists of each generation chose the systematic description and analysis of landscape and the processes that change them as specialised area.

We may know *geomorphology* in this way as *geo* = earth, *morph* = forms and *logy* = study (logos). In other way, geomorphology is the study of the forms of the earth.

In fact, it is not simply described by the heights of its hills and the slopes but by the reconstructed geological history of its evolution. It is not possible to understand the landscape until the entire geological history of the rocks and slopes is known. The area emerged from the "Cretaceous Sea". *Physiography* is often regarded as part of the larger science of geography but genetic geomorphology is mostly a geological subject.

According to Dury (1972), geomorphology is a part of two sciences and geomorphologists may be regarded as either geographers or geologists or preferably both.

Process oriented geomorphology is related to climatalogy, because air temperature, precipitation, winds and atmospheric humidity largely determine the response of rocks to sub-arial exposure. Geomorphology has evolved is summarized by the trinity of structure, process and time. The questions of what, where and how as in the following two points which are as under.

1. "The city of Ithaca, New York, is built on a delta plain at the south end of the Cayuga Lake through glacially oversteepened valley walls around the city are in contrast to the maturely dissected, crustaform upland of the glaciated northern Appalachian plateaus."

2. "The city of Ithaca, New York, is built on a low ground at the south end of the Cayuga Lake. Steep slopes surround the town, above which is a broad, rolling upland that becomes gradually more ragged southward into the low mountains of Pennsylvania."

The landform as the unit of systematic analysis. Landscapes are surfaces composed of an assemblage of subjectively defined, irregular, lesser surfaces.

We know that a landform is defined as each element of the landscape that can be observed in its entirety, and has consistence of form or regular change of form. Observer sees a hillside as landform gullies, ridges, and other small topographic features that either break the continuity of the hillside or establish its continuity, depending on the mental concept.

Because of the peculiar intensity of atmospheric weathering and erosion on our planet, most sub-arial landforms are destructional or erosional. Each landform can be visualized as enclosing or covering a mass of rock with specific physical and chemical properties and geometrically disposed discontinuities, such as bedding planes, joints faults, etc.

A major topic of geomorphic study is whether all stream valleys are always graded (fully adjusted to the erosional processes that are actively shaping them) or whether the graded condition develops slowly and progressively.

As we see the further considerations for structure, process and time. The minerals that form the rock the structural and statigraphic arrangement of rock layers and masses, the previous tectonic displacement of the roc! and the present state of deformation.

According to Carey, 1954, the property of a substance that determines whether it will behave as a solid or fluid has been defined as rheidity, and is arbitrarily measured by the amount of time necessary for viscous flow to exceed by 1000 times the elastic deformation, under specified conditions of temperature pressure and shear stress Fig. 2.1.

Gravity tends to pull down structures that rise above the general ground level. Geomorphic process also varies in intensity from one region to another.

Time is now specified to a precision that was not conceivable a few decades only. When this is applied to geomorphic problems,

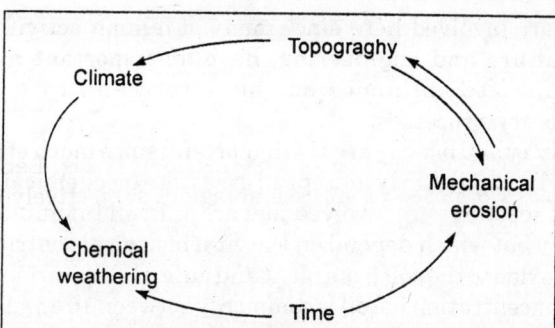

Fig. 2.1: *Cycle of erosion.*

radiometric ages can answer such questions as when an ice sheet uncovered a region; when rivers began to dissect a volcanic ash bed, etc.

A NEW APPROACH TO THE SCOPE OF GEOMORPHOLOGY

In fact, there is no neat framework into which landforms can be realistically placed. But there is little doubt about the scientific ingredients of a modern geomorphological explanation. These are drawn from the study of geomorphical process, a study which in turn depends on the working knowledge of a wide range of sciences. These include geology as the indispensable, basic science concerned with the properties of the raw materials of landform.

But to understand the geomorphological processes which sculpture rocks, transport debris and which modify geological structures, information is essential from other services, including climatology, hydrology, pedology and ecological studies. Furthermore, an appreciation of the physical motions of water, wind, ice and the nature of their role as transportation agents may be deepened by an understanding of physics.

Biophysical processes are also at work, shaping the landform around us. The rate of reworking of soils by animals has been known since the late nineteenth century to be impressive. In 1880, the English naturalist Charles Darwin calculated that earthworms annually bring to the ground surface about 25 tonnes of soil per hectare. Equally noteworthy is the varying effectiveness of a given vegetation cover in protecting soil form erosion. Even the social sciences are involved here since many of human activities, such as agriculture and engineering, have an important and often underestimated influence on the nature and rate of some contemporary processes.

Chemistry and biology are also important, since much of the rock-weathering, particularly in tropical areas, is a biochemical process. Chemical reactions are involved that are difficult to simulate in the laboratory but which depend on the vital biological contribution of carbon dioxide to the soil from plant and animal respiration. Carbon dioxide concentration in soil is commonly between 10 and 100 times greater than the 0.03 percent found in the atmosphere. The intensified carbonic acid dissolved in soil water accounts for the impressive solutional weathering associated with limestone.

At the outset, however, the influence of geology on landforms is fundamental. The disposition of rocks at the earth's movement and is expressed as structures like faults or folds. Moreover, the characteristic properties of rocks are important in relation to their resistance to erosion.

Controversy exists about the influence of climate on landforms, but some of the processess which lead to rock breakdown and the subsequent movement of the fragments or detritus are most distinctive in a given climatic zone. Particularly, intrusive in the context are studies of the processes operating in arid and semi-arid areas such as the waterless deserts of Africa and the inhospitable "badland" of North America. The scant vegetation cover in these areas leaves the actual shapes of landforms starkly obvious. Such areas also show clearly the physical breakdown of rocks and the implication of a spare vegetation cover when an occasional downpour of rain rests upon unprotected earth.

Water flowing down to the ocean over the land surface is the dominant agent of erosion. Erosion by stream action tends to broaden and deepen a valley floor, particularly if undercutting leads to land sliding.

Usually, water percolates vertically through soil, then down through the pores and joints of the underlying rock, where it increases the store of underground water. This water slowly continues to attack the deep rock, termed *deep weathering*, an action which in tropical areas may extend down to 100 m (over 300 feet) below the land surface. Such effects on landforms are most thorough by investigated in limestone terrain.

3 Basic Concepts and Significance of Geomorphology

The term *geomorphology* means a discourse on earth forms is the derivation of three Greek roots. Geomorphological studies encompass landforms of the continents, their margins and the sea-floor. In the study of landforms, the geologist looks into the geological controls in the evolution of landforms, the geographer is concerned with the adjustment of human activities in changing landforms and the engineer assesses the terrain from the point of view of engineering constructions and availability of materials for construction. In the study of landform by the geologist an in-depth understanding in the historical and dynamic elements of the process of change upon geological materials and structures, is of paramount importance.

As we know that all geologist and geomorphological activities of the earth are due to endogenetic and exogenetic sources of energy. Convection currents in the mantle and radioactive decay of minerals produce endogenetic energy which is manifested in the form of earthquakes, volcanoes, plate movements and so on. Solar radiation, gravitational attraction and biological processes are examples of sources of exogenetic energy fluvial, glacial, eolian erosion of crustal rocks are also weathering and mass wasting of rocks are also attributable to exogenetic energy.

Landforms evolution on the regionally extensive scale is understandably spread over a long span of time whereas a ripple or a mudflow may form in a few minutes. So, landform units are studied with reference to their magnitude in space and time. Evolution of continents, mountain chains and extensive valleys is spread over a period of millions of years. While during the fifteenth, sixteenth and the seventeenth centuries, evolution of landform features was based on the philosophy of catastrophism, the principle of uniformitarianism propounded by James Hutton (1726-1797), made a profound impact, paving way for scientific analysis of landform evolution.

Principle of uniformitarianism by James Huttn Conveys that the present is the key to the past, which is a fundamentals concept in geomorphological studies. Though Hutton applied the principle rigidly arguing that geological processes have been active at the same level of intesity throughout geological time, it is now recognised that it is not true. The common consensus is that the physical processes and laws which are in operation during the present time were active in the geological past also, but not necessarily with the same intensity as now. The second basic concept is that in the evaluation of landforms, structure plays a crucial role and is reflected in the landforms. The term structure as used here is a sack term including multifarious like joints, bedding planes, faults, folds, physical hazardness, constituent minerals, permeability, etc. Geomorphic features developed on rocks are, in general, much younger to the structural features of the rocks. The third concept is that geomorphic process operate not on a uniform rate but at differential rates resulting in the evolution of relief features. The processes are influenced by varied factors as altitude, topographic configuration, temperature, moisture, and the type and quantum of vegetation. Variations in microclimatic conditions also lead to subtle differences in landforms. The fourth concept envisages that geomorphic processes produce distinctive imprints on landforms, and characteristic assemblage of landforms evolve as a result to the operation of different geomorphic processes. For example, deltas and alluvial fans result by stream action whereas subterranean caves and sinkholes in limestone terrains are attributed to ground water action.

The morphogenetic classification of landforms takes into account geomorphic agents and processes which operate under different sets of climatic conditions.

W. M. Davis (1850–1934), thought that landforms display distinctive characteristics depending on the stage of their development. According to him, the evolutionary trend of landforms is through youth, maturity and old age stages. Though many geomorphologiste are not convinced of this concept, as a gross generalization it is recognised that an orderly sequence of landforms comes into being by the action of different erosional agents on the earth's surface. The sixth concept is that geomorphic evolution is one of complexity then simplicity. The topography of a single set of geomorphic processes or a cycle of development. Ever in a limited area of study, remnants of landforms not related to the current cycle of erosion are recognised. For example, in the deeply eroded Pre-Cambrian, while the majority of landforms may by attributed to the current cycle, remnants of landforms of even jurassic age are recognised. Five categories of landforms including simple, compound , morocyclic, multicyclic and exhumed have been recognised. While a single dominent geomorphic process results in a simple landscape, compound landscape bear evidences of the operation of more than one geomorphic process. It is true that all landscapes on the earth's face are compound in nature, but in the evolution of certain particular landscapes, the dominance of the operation of a particular process may be evident. For example, in the evolution of a delta, stream action is of primary importance though weathering and movement of material by force of gravity are also involved. Monocyclic landscapes resulting by the operation of a single cycle of erosion are rare in the major topographic features of the operation of more than one cycle of erosion. Polyclimatic landscapes evolve under more than one type of climatic environment, and change in the climatic pattern world naturally mean variation in geomorphic processes landforms formed in post-geological times and buried beneath sediments and igneous rocks may be laid bare by erosive agencies. In same cases, the existence of the surface, the backdrop of hills of the Tirupati University show the Cuddapah formations resting on Pre-Cambrian rocks. Here the Pre-Cuddapah landforms and blanketed by the Cuddapah sediments and thus a fossil surface is recognised.

As there are many concepts of fundamental nature, in which another concept is that little of the topography of earth is older than tertiary and most of the topography is no older than Pleistocene. A scientist, viz. Ashley (1931), estimated that nearly 90 percent of the land surface of the present day developed in post. Tertiary times and as much 99 percent is post–middle Miocene in age. The face of mother earth has been literally changing under the impact of the operation of exogenetic and endogenetic agencies. While the rocks from which the landforms are carved out may be ancient, the landforms themselves may be much younger. For example, while the evolution of the Himalayas dates back to early tertiary times, their present elevation was not attained till Pliocene and most of their topographic features are Pleistocene or younger. In Peninsular India, Pre-Cambrian crystalline rocks are exposed in force, but the landforms carved out of them are much younger. Similarly, the structures of the crystalline rocks are much older than the topographic features.

Still another concept of fundamental importance in geomorphic studies is that in the interpretation of the present day landform features the varied influences of climatic and geologic changes during the Pleistocene period are to be considered. Pleistocene glaciation was regionally extensive and its influence was far more widespread than the area covered by glaciers. The climatic effects of glaciation were well-defined and profound. Humid climate was prevalent in some areas now falling under the influence of temperate climate, permafrost conditions prevailed. Glaciers modified stream courses as exemplified by the Chios and Missouri rivers in the North American continent. Since ice was formed from water, sea-level was lowered by at least a hundred metres. The disgorging of cold glacial melt waters in to oceans, would have adversely affected reef building corals.

Geomorphic processes are influenced by climatic factors, particularly those of temperature and incidence of rains is well recognised. Infact, the study of paleo-climatic patterns gains importance in the study of the operation of geomorphic processes in the geologic past. The direct control is through the intensity and amount of precipitation, temperature fluctuations and so on. Indirectly, the kind of vegetation which is sustained by climate factors controls the operation of geomorphic processes. Another important concept of far reaching importance is that though the geomorphologist is concerned with the evolution of the landscape as at present, geomorphological studies are of maximum usefulness when extended back in time in the geological past. Scientific analysis of landform features in several parts of the world do indicate that remnants of features of the geological past are preserved.

In 1967, L.C. King, has forcefully brought out remnants of planation surfaces in the continents of the southern hemisphere and has correlated them.

4 Geomorphology and Environmental Hazards

Some of the hazards in our environment have geomorphological implications. Some hazards like hurricanes and blizzards belong to the realm of metorology, but then they may result in sea floods and avalanches which have geomorphological overtones. While hazards like volcanic activity and earthquakes are entirely natural disasters, a few are related to human interference with environment. The range of geomorphological hazards is large and include earthquakes, volcanic activity, destruction by weathering, fluvial erosion, slope instability, coastal events, periglacial and glacial destruction processes.

Human interference compounds soil erosion which is recognised to be a major disaster the world over. The two separate events which are involved in soil erosin are detachment of particles and their transportation, high rates of erosin have been recorded in areas with high seasonal rainfall and in semi-arid tracks. As a disaster soil erosion takes dynamic proportions under conditions of mismanagement of both arable and grazing lands; when forest fires and unimaginative tree felling take place; when vegetated lands are laid bare under conditions of urban development and laying of roads and when mining activities flourish on a large scale. Withal soil erosion is conditioned by multifareous factors including soil characteristics, topography climate vegetation cover and modes of land use. Soil erosion does not have repurcussions just in the areas of soil loss, but has extended implications. For example, soil washed into river channels leads to channel aggradations, flooding and deterioration of the quality of water.

Wind velocities are generally high in arid and semiarid areas, where soil erosion by the agency of wind needs to be tackled in an imaginative manner. Studies have been carried out by laboratory simulations of erosion process, monitoring of losses of soil in the field and by prediction of wind erosion. Population explosion in several countries has resulted in overgrazing by cattle, over-cultivation to meet food needs, over-pumping of ground water and so on. Appreciable decrease in rainfall due to removal of vegetation has been identified in several areas.

Right through history flooding has been a widespread natural disaster. Flood plains and alluvial fans have been prone to inundation. In many cases, the dimensions of disasters had been known in advance and to the cause identified. The most common causes of flooding are heavy rainfall and melting of glaciers. The other causes include volcanic eruptions beneath glaciers, draining of englacial and subglacial lakes and breaching of dams. Infiltration capacity of soils and moistures conditions influence coversion of precipitation into discharge. Land use changes can also cause flooding urbanisation leads to the construction of impermeable surfaces and gutters leading to gushing of surface waters.

Slope instability leads to simple soil creep of little consequence and also disastrous landslides. Mudflows, volcanic mudflows (Lahars) and bog-bursts are known to lead to catastrophic consequences. Simply put, landslides are triggre when the forces responsible for movement exceed those resisting it. In essence, movement takes place from an unstable to a stable state. Man has contributed his share to slope instability by modifying natural slopes.

Subsidence of ground surface is a problem to reckon with in the present stage of industrial development withdrawal of fluids including water and oil form depth leads to slow but sure and extensive subsidence. Compaction of sediments after irrigation and land drainage as also mining of solid ores cause subsidence. In the Jharia Coalfield in Jharkhand, extensive subsidence of land due to removal of coal from depth has been known for long. Underground fires in coalfields is yet another problem. Management of environment to prevent and retard subsidence has received due attention in several countries.

Geomorphological process act with rapidly in the coastal environment. Flooding, erosion, coastal deposition and pollution are the factors involved. Long-term, retreat and erosion of the coast, short-term erosion and degradation of beaches are world-wide phenomenon. Coastal erosion is a major problem along the west coast of India. Flooding of sea waters may be due to hurricanes, tsunamis and storm surges. Depressions in the Bay of Bengal are known to wreck havoc along the eastern coastal stretches of India. As for coastal deposition, silting of harbours and parts is a well-defined problem in India. The pollution problem is hydraheaded and is receiving increased attention because of the ever-increasing threat posed by the discharge of effluents into rivers and seas.

In periglacial environment, permafrost frost heave, cracking of frost and thawing resulting in slope instability are identifiable hazards. Man has made inroads into not only deserts and tropical rain forests but also into periglacial environments, where vegetation has been removed, drainage conditions modified and brick and mortar structures built. Such interference has degraded regions of permafrost and disrupted the thermal equilibrium of the ground.

Rivers are generally get blocked by avalanches in glaciated mountain areas which cause floods with catastrophic consequences. Research activities are a foot in Switzerland to mitigate the destructive consequences of avalanches. Stabilising covers of coniferous forests and construction of vaults, bunkers, galleries and the like have been adopted.

Volcanoes and earthquakes, though strictly geological topics have been studied by geomorphologists also. Though in general areas of volcanic activity are thinly populated some island-arc areas as Java and Japan are exceptionally densely populated. The hazards posed by volcanoes include lavas, tephra ejected by volcanoes, gases and nuees, aredentes as also side effects like heavy rains, mudflows and landslides. The enormity of damage caused to life and property by earthquakes bears no repetition. In the earthquakes that rocked Armenia in December, 1988 more than a lakh of lives were lost. Earth movements, shock waves, subsidence of ground, disruption of ground water, creation of tsunamis are the direct consequences of earthquakes. Man's response to earthquakes and volcanic activity is focused on understanding them in their true perspective so that predication and safety measures can be adopted.

5 Geomorphic Processes and Parameters

In fact, physical and chemical changes bring about modifications of the surficial features of the earth are known as gemorphic processes and any natural medium which by its innate capacity acquires and transports materials is a geomorphic agent.

Exogenetic agencies are those which originate outside the earth's crust, whereas those emanating from the bowels of the earth are endogenetic. A broad classification of the processes is as follows.

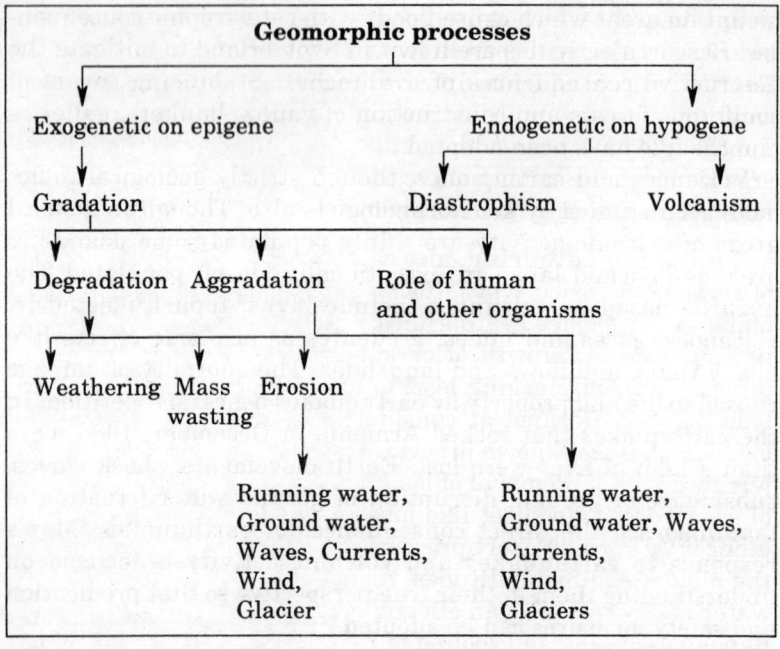

Excluding detailed above, the impact of meteorites with the earth, modifying the earth's crust is recognised as an extraterrestrial geomorphic process.

Gradation is considered to include in its fold all the process which tend to bring the earth's crustal features to a common level and the processes involved are those of degradation and aggradation.

WEATHERING

Weathering is the mechanical fracturing and chemical decomposition of rocks *in situ*, by natural agents at the surface of the earth. In other words, weathering which in essence is the disintegration or decomposition of rocks *in situ*, includes a number of processes which by concerted action reduce gigantic rock masses to clastics. The factors which have variable influence on weathering are the structure of the rocks, the topography and vegetation of the terrain and the prevailing climate. Structure of the rock includes a number of physical and chemical characteristics. While physical characteristics like bedding plane and fracture determine the ease with which water enters into rocks, chemical characteristics have bearing on the susceptibility of the minerals to weathering. Temperature and humidity have great bearing on chemical weathering as exemplified in tropical and sub-tropical areas. Topography determines the extent of rock exposures. Decaying organic matter resulting from vegetal cover contributes humic acids and carbon dioxide which are agents of chemical weathering.

Physical weathering, also called **mechanical weathering** of rocks is of varied types. Rocks expand due unloading of rock masses and repeated heating and cooling arising out of fluctuations in temperature, growth of crystals, activity of organisms and probably colloid plucking lead to physical weathering. Relief of load on rocks formed at considerable depth may give rise to exfoliation domes known in several localities in the Pre-Cambrian terrain of India. Removal of load by erosion may lead to splitting of sandstones into sheet-like masses, growth of ice crystals and other crystals may exert pressure leading to rock fracturing. In the middle and high latitudes, repeated freezing and thawing of water is recognised as an effective agent of physical weathering. Repeated heating and cooling of rocks, especially in deserts may produce strain in thin layers of rocks masses leading to exfoliation.

Different minerals in rocks have varied coefficients of expansion and rocks are poor conductors of heat. The origin of some spheroidal boulders, formerly attributed to alternate heating and cooling is now considered to have been largely due to effect of oxidation and hydration of minerals. Growing plants penetrating into rock crevasses and fractures may well widen them. Decaying animal and plant remains may form carbon dioxide and organic acids which may act on rocks, resulting in chemical weathering. Colloid plucking, attributed to plucking small bits of rocks by soil colloids is a minor agent of mechanical weathering.

Main chemical weathering processes are the result of hydration, hydrolysis, oxidation, carbonation and solution. Crustal rocks exposed on the surface of the earth and subject to chemical weathering yield products which may be aptly described as products of sub-aerial, residual, physicochemical weathering. In general, chemical weathering is more active in tropical and sub-tropical regions receiving good incidence of rains. In arid regions also chemical weathering is recognised, though advanced typed of chemical weathering does not take place. Chemical weathering of rocks may lead to the formation of minerals with density lower than the primary minerals present in the rocks; increase the bulk of the rocks leading to the building up of strain in the rock masses; produce smaller particle size which would understandably increase the surface area or interface between the constituent particles and the media surrounding them; result in the formation of mobile materials; and also may give rise to more stable minerals in consonance with the prevailing environment. The increase in the interface area as a result of production of smaller particle size naturally presents a large total area for the operation of the process of chemical weathering.

Mineral stability under changing environments is a matter of importance in chemical weathering. Consider the mineral assemblage of a granulites like charnockite, extensively recognised in the Indian Peninsula. The rock was formed at deep tectonic levels with attendant high pressure and temperature. In essence, the minerals of the rock attained equilibrium under high pressure and temperature. The minerals of the rock when exposed at the surface level are in an environment entirely different and so are in a metastable conditions. Chemical weathering of the minerals tends to produce minerals which are at equilibrium under subaerial conditions. The following table of Goldich (1938) gives the order of increasing stability of minerals, from top to bottom.

Olivine	Calcic plagioclase
Augite	Calc-alkali plagioclase
Hornblende	Alkali-Calc plagioclase
Biotite	Alkali plagioclase
	Potash felspar
	Muscovite
	Quartz

(↑ Increasing stability, at left margin)

The above table of Goldich has a distinct similarity to Bowen's Reaction Series though the subjects dealt with by the two are distinctly dissimilar. It is necessary to bear in mind that the Goldich's table just bears out of the degree of increasing stability from top to bottom. For example, it is not to be inferred in comparision with Bowen's table that augite weathers into hornblende and hornblende to biotite. On the other hand, in rocks composed of olivine, hornblende and augite, olivine weathers more readily than augite or hornblende. The minerals at the bottom of the table, muscovite and quartz, commonly present in igneous and metamorphic rocks are more or less in equilibrium under conditions obtaining on the earth's surface. In laterites representing tropical weathering profiles, quartz is often preserved as discrete grains.

Pocesses of chemical weathering are inter-related under natural conditions. The process of hydration involving in fact both hydration and hydrolysis is one of absorption of water as for example, the formation of gypsum from anhydrite as shown in the following formula.

$$CaSO_4 + 2H_2O \longrightarrow CaSO_4 \cdot 2H_2O$$

Likewise the hydrated oxide, limonite, forms from hematite.

$$2Fe_2O_3 + 3H_2O \longrightarrow 2Fe_2O_3 \cdot 3H_2O$$

In the above two, no chemical change is involved and the process are reversible.

Hydrolysis, involving formation of hydroxyl is well-observed in the weathering of micas and felspars as denoted by the equation.

$$KAlSi_3O_8 + H_2O \longrightarrow HAlSi_3O_8 + KOH$$
(Feldspar)

The aluminosilicic acid, being unstable gives rise to colloidal silica and a colloidal complex. The hydroxide (KOH) will form potassium carbonate by reacting with carbon dioxide as given by the equation.

$$2KOH + H_2CO_3 \longrightarrow K_2CO_3 + 2H_2O$$

The potassium carbonate, thus formed by carbonation is apt to be carried away in solution since it is water soluable.

In the process of weathering of feldspars, hydrolysis and carbonation tend to produce a water soluble carbonate as shown by the equation.

$$2KAlSi_3O_8 + 2H_2O + CO_2 \longrightarrow H_4Al_2Si_2O_9 + K_2CO_3 + 4SiO_2$$

In chemical weathering hydrolysis precedes oxidation though the process of oxidation is fast in sulphides and carbonates and silicate minerals are also affected. For example, hydrolysis of olivine results in the formation of magnesium, hydroxide, silicic acid and ferrous oxide as indicated by the formula.

$$4FeO + 3H_2O + O_2 \longrightarrow 2Fe_2O_3 \cdot 3H_2O$$

Calcium corbonate is easily taken into solution, the reaction being as follows:

$$CaCO_3 + H_2O + CO_2 \longrightarrow Ca\,(HCO_3)_2$$

In this reaction, CO_2 is contributed by decaying organic matter.

The common rock forming minerals like micas, olivines, feldspars, etc. when subjected to hydrolysis and oxidation, show an increase in bulk, become soft and loose their elasticity and lustre. They are ready to be acted upon by physical and chemical weathering processes.

The pH of ground water has over-riding influence on rock weathering. In the case of silica (SiO_2), it has been found that it is moderately soluble throughout the common range of soil pH and the solubility rises steeply under alkaline conditions, with pH values higher than seven. Alumina (Al_2O_3) is also moderately soluble throughout the common range of pH and there is increase in solubility under alkaline conditions. Under conditions of pH4 and less, alumina is soluble, and under the same conditions silica is much less soluble. However, highly acidic environments with pH4 and less are uncommon in natural setting. Iron in reduced conditions as FeO is mobile under acidic on neutral conditions, while under alkaline conditions, it is precipitated.

The zone of weathering of the earth's crust is that portion of the lithosphere which has subjected to weathering processes under sub-aerial conditions. Since climate, especially temperature and moisture, have profound influence on chemical breakdown of minerals, the weathering styles as also depth of weathering vary from one climate zone to another.

Mass Wasting

The various types of mass wasting have been classified by Sharpe (1938) under four heads, including slow flowage type, rapid flowage type, landslides and subsidence. The causes which activate the processes may be passive or active. Passive causes may be due to weak and loose consolidated materials turning slippery when wet: presence of thinly bedded and alternating permeable and impermeable beds; presence of weak zones like faults, joints and so on; steep ground slopes; large changes in diurnal and annual temperature ranges; and poor or no vegetal cover. The causes which activate the processes may be due to steepening of slopes by running water, overloading by water saturation and excavation of material by natural or man-made agencies.

Crustal Erosion and Material Transportation

Each of the agents of erosion, including running water, ground water, waves currents, wind and glaciers has a distinct style of processes. However, erosion encompasses, acquisition of loose material, grinding and wearing down of the bed rock by the material that is in movement, mutual attrition of particles in movement and transportation of the basis. Solution may be an important aspect under some conditions.

Aggradation

When weathering of rocks occur, the resulting debris is deposited and agents like wind, running water, ground water, currents and waves play their part in depositional processes. No doubt, sculptured landforms produced by erosion are more attractive than depositional landforms, but then both are of equal importance in the study of features of the earth's surface.

Endogenetic or Hypogene Processes

In fact, there are nearly 60,000 volcanoes on the earth, in which, about 50,000 volcanoes occur as seamount in the Pacific ocean floor. Accordingly, it is obvious that the volcanoes have an important role in modifying earth's surface. Upwelling of magma through the vent type and the fissure type volcanoes, naturally modify the topography. The gently undulating type of topography in the basaltic terrain in western India is due to the flowage of magmetic material. Understandably, extensive flows result in the formation of plateau landforms. The crustal rocks may also be deformed by the intrusion of magma, producing domal structures.

Distrophism involves orogenic processes marked by deformation of the earth's crust and epeirogenic processes which result in regional uplift of the crust without large scale deformation. Presently, it is considered that many orogens have a complex history extending over millions of years, including rapid and slow deformation of crustal rocks. In contrast to orogenic deformation, epeirogenic movements, involving vertical movements of the crust do not involve severe deformation of the crust. Epeirogenic movements active over a protracted span of time may cause the submergence of cratons by several thousand metres.

Perhaps, the most unusual landforms carved from the earth's crust are produced by the impact of meteorites the earth is known to colloid with large bodies known as Apollo objects but the frequency is fortunately low, being once in several million years. The Apollo objects are known to be interplanetary bodies, the orbits of which cross that of the earth. About 1200 Apollo objects with diameters of more than a kilometre are recognised. On the face of the earth about 100. Apollo craters with a diameter of more than a kilometre are known. However, the operation of weathering and erosional processes modify and destroy the craters. Craters are known to be better preserved on the lunar surface.

Human activity is also recognised to modify the earth's surface. Large quarries, rock cuts and fills, and other excavations are some examples. Winning of minerals for industrial utilisation also means modification of the total budget of the earth. It has been computed that nearly 85 percent of the land surface of the earth does not remain in its natural state. Apart from humans, termites and some birds and animals locally modify the earths's surface.

6 Geomorphic Cycles and their Interpretation

Willium Horris Devis (1850–1934) gave the concept of cycle in geomorphological studies. Davis was an American geologist, geographer and teacher. His familiar phrase. "Landforms are a function of structure, process and stage" laid the cornor-stone for conserted studies on the evolution of landforms. His evolutionary sequence of envents, including youth, maturity and old age stages in a cyclic form leading to the formation of a surface of low relief (peneplain) is not wholly acceptable to all geomorphologists.

However, the terms geomorphic cycle and cycle have been found useful in discussing the evolution of landforms. The term as they are used now, do imply and orderly and sequential development of landforms but do not indicate that landform evolution is strictly cyclic in nature and that the stages of a cyclic represent readily recognisable stages of landform evolution.

Again Davis says, a surface rapidly uplifted is subjected to prolonged erosion under tectonically stable conditions, the evolving landscape, passes through the stages of youth, maturity and old age, each of which has distinctive and recognisable characteristics; and finally end product is a peneplain, resembling the original surface before uplift. Thus, a cycle is run. Davis recognised that if the uplift was not rapid, then uplift was accompanied by erosion as well. He argued that many landscapes are polycyclic in origin and the erosive process aim at reducing the landscapes to the lease level of erosion. In the operation of cycles, volcanicity, glaciation and aridity were considered as accidents by Davis. The Davis had recognised that evolution of landscapes was a long-drawn process and so used geological time scales in his discussions.

A diagrammatic sketch, illustrating Davis' concept is given in Fig. 6.1.

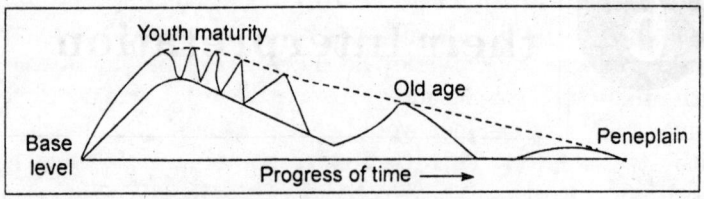

Fig. 6.1: *Davisian cycle of erosion.*

Several severe criticisms have been levelled against Davis' works. His greater emphasis on stage than on structure damaged his statement that "landforms are a function of structure, process and stage". He is also criticised for not studying the mechanics and the nature of the processes in the field. Davis' assumption of rapid uplift of terrains initially is considered unsatisfactory. It is doubled by many as a whether the youth maturity old age sequence and the landforms associated with each of the stages actually occur. In fact many landscapes display both maturity an old age characteristics. Some doubt if peneplains as envisaged by Davis can develop, since in the later stages of landscapes evolution, the slope is so subdued that capacity for erosion will taper off. Some consider that Davis impressions of the forms of slopes were in accurate and misleading. Above all, his work is considered to be based on deductive approach which is unscientific. Despite all the criticism, it is evident that the concept of cycle by Davis has endured for long and provided background for scientific enquiry concerning the evolution of landscape. Davis' concept with modification was applied in the analysis of the arid cycle, mountain glaciation, karst erosion, marine erosoin, peneplantion, savanna erosion, pediplanation and periglaciation erosion.

Walther Penck, a German geologist, was a severe critic of the concept of Davis, since the two viewed geomorphology from different angles. While Davis made explanatory description of landscape, Penck considered that the main thrust of geomorphological studies should be aimed at understanding earth movements. While Davis' model was based on rapid uplift followed by erosion under tectonically stable conditions, Penck visualised uplift over a long span of time with attendant erosion. Penck held the view that the stage of the landscape was determined by the rate of uplift of the land. Based on this assumption three alternatives of landscapes evolution were derived. Firstly, under conditions of constant rate of uplift, a constant rate of down cutting by rivers would produce straight valley side slopes, resulting in a landscape with medium relief. Secondly, if the rate uplift was accelerated, convex valley side slopes would evolve, resulting in a landscape with strong relief. Thirdly, if the rate of uplift declared the resulting slopes would have concave structure with the cessation of uplift and continuation of erosion, a gently undulating surface, called *terminal surface,* would come into being Penck's views have been criticised as were those of Davis, his idea that slope and gradient are determined primarily by rates of erosion by rivers is questionable. Penck regarded structure, vegetation, weathering and transport of wethered materials as of subsidiary importance but modern geomorphological research has highlighted their importance in the evolution of landscapes. However, it is to the credit of Penck that his concept of parallel slope retreated and slope replacement is increasingly considered important.

In 1950, L.C. Kind worked out a new cycle of erosion to explain the erosion surfaces of Africa, where two main elements of the landscape needed to be analysed.

The first of these elements was the erosional scarp cut into bed rock (pediment) showing a concave profile with low slope angles (up to about 1°) and the second, the scarp with a slope of 15–30°. The pedement and the scarp were separated by a topographic break. King visualised the recession of scarps without change of angle and called the process, "Scarp retreat". In general, the uplift of

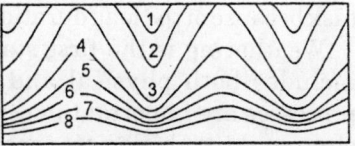

Fig. 6.2: *The uplifted land subjected to erosional cycle is represented in the three stages of erosion: Youth–1-2; Initial maturity–3; Middle maturity–4-5; Final maturity–6; Old age–7-8.*

a plain was followed by dissection and the scarp slopes retreated paralled to themselves. The process resulted in formation and enlargment of pediments; and when the hills are completely worn out by back wearing, the resulting landscape made up of coalescing pediments has low relief (Figs 6.2 and 6.3).

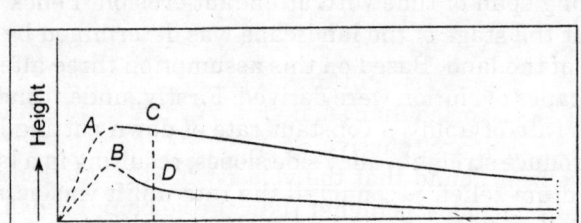

Fig. 6.3: *The geographical cycle of Davis.*

King identified several erosion cycles in South Africa and brought out with force the coexistence of stepped planation surfaces. His insistense on back-wearing of slopes rather than down wearing meant that remnants of high level erosion surfaces might be preserved. He made an intercontinental correlation of the erosion surfaces in the southern continents and identified the "Gondwana Surface" of Jurassic age in these continents.

Historical geomorphology was more of an explanatory science than a descriptive one. Modern geomorphology puts greater emphasis on active processed and resulting forms. In fact in modern textbooks on geomorphology the process form relationship is highlighted. Study of processes in the field is by no means an easy task because of the long duration and slow intensity of the processes.

Weathering is the first stage of the cycle of processes, which includes both physical and chemical weathering. Biological processes are important in certain environments and it has been observerd that biological processes may retard or accelerate both physical and chemical weathering of crustal rocks. Soils which are products of weathering are related to biological processes and constitute a zone between sub-aerial processes and the bedrock. The movement of soils in relation to slopes is important in geomorphical research. The "Catena Concept" brings out the close relationship between soils and landforms. The processes under erosion have long influenced studies in geomorphology. Rivers erode by abrasion, corrosion ad hydraulic action as also by the addition of particles to suspended load in sediments. Glacial processes include plucking and abrasion is laid on chemical weathering. The subject of transport of debris has gained much importance because of its practical implications, as for example, soil erosion by water and wind. Two points which have come to force on the basis of modern studies on transport are:

(a) Sand with grain size in the range of 0.06–2.0 mm is more easily transported than smaller and larger sized grains. Fine salt and clay are cohesive.

(b) It has been found that the velocity of currents required to entrain particles, is higher than the velocity needed for the transport of particles.

It is held that the landscape is all slopes and accordingly, slope form and processes giving rise to slopes are important in modern geomorphology. The subject is one of practical importance since ground subsidence and slope in stability are covered by the slope element. In the classification of slopes, two main processes including action of water and mass movements are considered. Action of water involves splash erosion resulting from raindrop impact, overland flow and subsurface flow of water. Shear stresses on slopes give rise to mass movements and the movements are classified into flows, slides and falls.

The pioneers in geomorphology perhaps gave greater attention to erosional landforms then to depositing landforms. The imbalance is corrected by the modern trend in geomorphology studies. While the evolving of a depositional landforms in the landscape can be analysed, in the study of the evolution of erosional landforms, the process need to be inferred on the basis of the form of the erosional features. Besides, depositional landforms are to a large extent amenable to dating by modern tools of radiometric dating methods.

Above all depositional studies are significant in applied geomorphology.

7   Relief Features
Topography and its Relation to Structures and Lithology

"All faces of places and their forms decay, and that is solid earth that once was sea. Seas in their turn, retreating from their shore, make solid land, what ocean was before."

Ovid, Metamorphosis XV

Various type of landforms of the earth's crust may be broadly classified under three orders of magnitude. The first order comprises continents and oceans which are the largest features of the earth. The configurations of the continents and the ocean basins have not remained static. The shapes and the spatial distribution of continents in the present day are considered to be the result of the disruption of a super continent in the Mesozoic Era. Seafloor spreading, the evidences for which have been assiduously collected in recent years, is known to have significantly altered the topography of the sea-floor. In fact, it is considered that the chequered history of the earth ever since it was born at about 4,600 million years ago has been punctured by repeated breaking up and assembly of continents. Many ancient features have understandably dissolved in the mists of time. However, geoscientists have pieced together evidence which show that parts of the earth's crust had been elevated and depressed, extensive stretches of dry lands of the present day were under water, violent episodes of volcanic activity and impact of extra-terrestrial bodies has scarred the earth's face and mountain ranges had been heaved up and worn down. In the midst of all these variable factors, an orderly system of the growth of continents, marked by the definition of ancient crustal rocks (cratons) bordered by stretches of mobile belts is deciphered.

If we consider the slope element of the earth as a whole it is seen that the abysmal deep sea platform of the oceans, slopes upwards through the continental slope and the continental shelf to continental platform above which the high mountain ranges of the continents are defined Fig. 7.1.

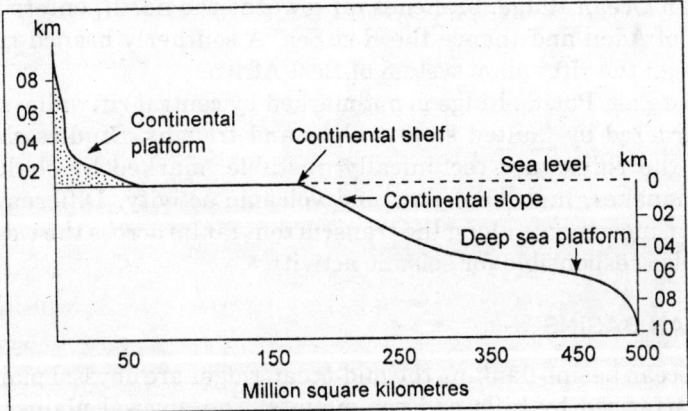

Fig. 7.1: *Hypsographic curve.*

Topographically, the earth's surface is inhomogeneous, while the oceans are spread over 70.8 percent of the earth's surface, lands constitute 29.2 percent. It is understandable that land and sea are mostly antipodal arranged. Only 1.5 percent of the surface has land antipodal to land. It is also known that about two-thirds of land is in the northern hemisphere. Contrary to popular opinion the deepest parts of oceans are not always far out from the land and are often located close to mountain ranges as in island arcs. While a medley of features are recognised on the earth. There are eight major topographic elements are significant which are as under.

OCEAN RIDGES

Wide oceans has brought to light traversal fractured linear ridges extending over a distance of about 64,000 kilometres. Studies have borne out that the ridges have widths of 2000–4000 km and rise from one to three kilometres or more from the ocean floor. The ridges in the Atlantic and Indian Oceans are irregular the East

Pacific Ridge is a smooth arch. The Mid-Atlantic Ridge is characterised by 25–50 km wide axial rift valleys, flanked by submarine ridges. Huge transform or transcurrent faults are defined more or less at right angles to the axes of the ridges. The emergence of the Mid-Atlantic Ridge above the sea level to form the island of iceland is a notable feature. The Carlsberg Ridge, a branch of the Indian Ocean Ridge, branches off towards the north, enters the Gulf of Aden and thence the Red Sea. A southerly branch runs through the rift valley system of East Africa.

The East Pacific Ridge is not marked by central rift valleys. It is bordered by faulted steps, ridges and troughs. Studies show that the ridges are tectonically unstable, marked by shallow earthquakes, high heat flow and volcanic activity. Differential lateral movements along the transcurrent faults across the ridges are also responsible for seismic activity.

OCEAN BASINS

The ocean basins flanking the mid-ocean ridges are abyssal plains, characterised by hills and sea-mounts. The abyssal plains are tectonically inactive and have gentle gradients of less than 1: 1000. They are known to be up to 1,000 kilometres in width below water columns measuring 3–6 kilometres in depth. As for sediments, they are found to increase in thickness towards the continental slope and shelf. The sea-mount in the bed are spectacular features with width of 2–100 kilometres and rising to dizzy heights of more than a thousand metres from the abyssal plains. The Azores and Hawaii islands are volcanic islands whose origin is attributed and thermal plume activity. Sea mounts have both sharply pointed and flat tops and they are in the form of hills whose tops are in some cases below a water column of 200 metres. Steep-sided seamounts (12°–35°) are known as *Guyots* and they appear to have ware leveled platforms whose submergence is attributed to sea-floor subsidence and rise in water level in post-glacial times. The growth of corals which are built in tropical waters at depths not exceeding 150 metres supports the view.

CONTINENTAL SLOPE AND RISE

In fact, the present shoreline of ocean does not limit the extent of continental rocks. The fact is also that the outer edge of the continental shelf, approximately located 0.135 kilometres below the sea-level delimits the continental rocks. The continental slope from its outer edge descends at slopes up to 6° to depths of two kilometres. Continental sediments in the form of coalescing fans and aprons mark the base of the continental slope. In the seaward extension of large rivers submarine canyons mark the continental shelf and slope off the east coast of Australia, the shelf is characterised by thick coralline limestones.

OCEAN TRENCHES, DEEPS OR TROUGHS

The Western Pacific and the Caribbean Coastal area, the continental shelf plunges steeply into trenches, some of which define the deepest parts of the ocean floor. They are known to be variable in length (300–5000 kilometres) and 30–100 kilometres in width, with slopes of 10°–16° in their deeper parts. The trenches run parallel to island arcs or young volcanic zones on their seaward side. A characteristics feature of trenches is their negative gravity anomaly.

Island Arcs

Most ocean trenches on their landward side are marked by parallel accurate festoons of islands as in New Guinea and Hikkaido. In certain cases they are topographically and structurally continuous with continents belts of young folded mountains such as in Malaysia, Alaska and Kamchatka. They are tectonically active zones with profound seismicity. While they slow strong negative anomalies, positive anomalies are marked on their continental sides.

Marginal Sea Basins

Marginal sea basins occur between island arcs and continents as for example the sea of Japan and the sea of Okhotsk, some of them are 500 to 1,000 kilometres wide and have rugged bottoms with faults, undulations and small sea-mounts, characterising complex histories and different sediment sources. Both tectonically active and inactive basins and known.

Folded Mountains

Sediments under the impact of compression. Folding, thrusting and uplift are thrown up as curvilinear mountain chains which may be associated with volcanic activity, deep igneous emplacement and metamorphism. Some of the young fold mountains are continuous into island arcs.

Broadly, fold mountains may be divided into older and younger groups. Those of the older group have medium scale elevations and are tectonically more stable than those of the younger group. The younger group includes mountains of the highest terrestrial elevations like the ALPS, the Himalayas, the Andes, the Rockies and the Pyrenees within these, large area of undeformed rocks may present as in the mountains of the younger group is borne out by seismically active belts and strong magnetic and gravity anomalies.

Relief features of the second order comprise mountains and plains which have resulted by the action of the internal forces of the earth. Such action of forces emanating from deep within the bowels of the earth includes both orogenic and epeirogenic movements. Mountain building movements by the arching of sediments produce dome-shaped mountains and block mountains by dislocation and elevation. Epeirogenic movements over a protracted span of time lead to crustal arching with little evidences of tectonic disturbances. The third order features result by the action of destructional force and give rise to residual features of peaks, erosional features including valleys and canyons and depositional features like deltas and glacial morains. In essence aided by weathering, streams, waves, wind and glaciers produce relief features of the third order.

Topography and its Relation to Structure and Lithology

Crustal rocks get curved resulting land features which have distinctive lithology and structure. The features represents a complex response to primary rock structures, secondary structures and to the effects of exogenic forces inexorably acting on the crust. A variety of variations in landform features is recognised as due to lithological variations in landform features is recognised as due to lithological variations which may be evident over a few square metres to hundreds of square kilometres.

The uplift of a canyon or a valley leads to rejuvenation of streams and erosion gives rise to the formation of steep and even vertical sides. If the formations are of a sequence of shales and sandstones, the shales being softer are apt to be eroded more quickly and the comparatively more resistant sandstones may form ledges. In the retreat of the walls, the processes of erosion are focused on near the weak bedding planes between the two rock types. The terraces and structural benches in the Grand Canyon of Colorado, USA, have resulted by this process. Concerted action of the in exorable processes of erosion may ultimately lead to the formation of striped plains with isolated residual buttes and mesas.

The retreat of seas from homoclinal structures with sequences of sedimentary rocks having a general dip in one direction may lead to the extension of consequent rivers and their tributories. Fluviatile action leads to the erosion of less resistant rocks giving rise to cuestas with gentle seaward sloping dip slopes. The extension of subsequent steams allows resequent and obsequent steams to develop. Eventually consequent rivers may be beheaded leaving a wind gap, resulting in the disappearance of drainage. It has been observed that captures occur wherever rocks have headward cutting advantage due to rock slope or lack or rock resistance.

Folds formed at the depths of the earth are exposed at the surface by erosion. The crests or anticlines may be breached by prolonged erosion. Most fold mountains except the high ranges of tertiary thrust mountains are known for complex erosinal histories. For an area of simple folds, drainage and erosion in the initial stage lead to the shaping of anticlinal mountain and synclinal valleys. At a later stage, the development of lateral consequent erosion results in the breaching of anticlinal crests and cores. Deep and sustained erosion by subsequent streams results in a complex drainage pattern and in the end, inversion of relief marked by anticlinal valleys and synclinal hills, results.

Woolidge and Morgen (1959), demonstrated the inersion of relief features in an area where rocks are thrown into parallel folds. The completion of a second cycle of erosion in an area with inversion of relief may bring into being anticlinal hills and synformal valleys, resembling the original. It is of interest to note that the deeply eroded pre-cambrain terrain of South India, inversion of relief typified by synclinal hills and anticlinal valleys is known.

When earth's crustal rocks which are brittle in nature yield to stresses, strain fractures and faults result. It faulting is considerable and erosion negligible, a monoclinal flexure may result. The flexure may break giving rise to a fault scarp in which the height of the scarp represents the throw. In an intermediate situation in which erosion and uplift occur together, the rising block is eroded by headward valley cutting at the margins on which alluvial fans are built. Erosion, if continued for long may produce training speer facets on the fault scarp. In the third condition in which erosion exceeds the uplift rate of the upthrown block, a general flattened surface across the fault zone may come into existence. If there is stability along a fault plane erosion may excavate the fault to produce a fault-line scarp. Block faulting gives rise to upstanding fault blocks, defining plateau landforms. The Hercynian massifs of Europe and block mountains, also known as *horsts*. If fault blocks are depressed in comparison to their surroundings, a situation in which sediments transported from adjoining uplands are deposited in the trough arises. Rift valleys and fault grabens are known for sedimentary fills.

There are large numbers of variations is the shapes and sizes of landforms are attributable to lothological controls. These variations are in association with rocks over limited areas as in an outcrop as also with uniform rock types extending over regionally extensive terrains. The climatic environment in which erosion occurs also has strong influence on the resulting landforms. Features such as jointing, bedding, texture, permeability and so on also have influence on the evolution of landforms.

Those arenaceous rocks, which have slilica cement are resistant to erosion and areas under lain by such rocks have low drainage density and well-defined relief. Compared to silica, iron oxide is a much less resistant cement and in bedded or jointed rocks weathering and erosion are at a rapid pace. Landforms on clay and shale have low relief as on the terrains of Talchir Shales in Indian Coalfields. In humid regions argillaceous terrains have low slopes ($\angle 8°$) and moderate drainage densities. Generally, dendritic drainage, controlled by regional gradients prevails. In sub-humid regions, with lack of vegetation and fairly high rates of evaporation, "Bad Land" topography with low relief and high drainage density comes into being.

The susceptibility of calcareous rocks to solution produces distinctive landform assemblages. The karst (meaning bare stony ground) terrain in areas underlain by calcareous rocks is characterised by the development of underground solution networks and the surface in pitted with holes and depressions. In the classical karst terrain of Yogoslavia, considerable thickness of pure limestone, elevation of the terrain above sea-level and high annual precipitation have together given rise to a unique topography. In the development of the karst topography, both surface ane subsurface waters have crucial roles to play. The Swiss Alps are known for the development of underground care systems. Many care systems as for example, the ones in Kentucky, USA, have long explored cave passages. The Gouffre Berger cave in France goes down to a depth of about 1100 metres below the ground level. In some karst terrains, closed depressios with depths of more than a hundred metres called *dolines*, are know. In hungry, large depressions and dolines in karsted triassic limestone are found to contain bauxite deposits.

Igneous rocks are resistant to erosion and the landforms produced reflect the original geometry of the rocks. The geometry of volcanic landforms depends on the type of magma, the morphology of the vent through which magma wells up and the denudational history of the terrain. Flat-topped plateaus form by the flowage of basalts as in western India and the setting down of volcanic ash. The ignimbrites of New Zealand are marked by flat-topped plateaus. Erosion working around lava plateau margins may produce *mesas* and *buttes*. Scoria and ash cover are generally not very much affected by weathering because of their high permeability and relatively small catchments area. However, relentless erosion produces a ribbed surface with radiating gullies. A high degree of erosion is known in large stratovolcanoes. Erosion of large domes, guided by foliation and jointing is particularly known in sub-humid climatic regions. Jointed crystalline rocks in tropical areas are known to be weathered to depths exceeding a hundred metres. Weathering along joint planes produces concentrically weathered core stones. The "Weathering Front" showing successive limits of weathering with the passage of time may be made out in some localities. In the residual profiles derived from crystalline rocks exposed in the west coast of India rounded core stones are seen within weathered rocks and lithomarge. Here the rocks are subject to inexorable chemical weathering under the impact of monsoon rains and prevalence of high temperatures throughout the year.

Mineralogical composition and structural elements like joints and fractures influence the weathering of metamorphic rocks. Quartzite with well knit interlocking quartz grains is resistant to mechanical and chemical weathering and so forms high relief features. Slates are susceptible to weathering and erosion and so present a subdued relief. Gneissic rocks behave like granites in general but layered mica is susceptible to be eroded and in the long process may give rise to rubble and sand.

<table>
<tr><td></td><td></td></tr>
</table>

8 Major Landforms

Wylie (1976), has classified the major features of the solid earth which are as follows.

Major subdivisions	Surface area % of total
1. **Continent and Continental Shelf Continental Slope**	40
Ocean Basin	60
2. **Active belts, traversing ocean and continents**	
Ocean ridge and rise	23.2
* *Mountain Ranges*	
Active, younger than 65 million years	2.8
Older than 65 million years	7.5
* **Volcanic Islands Area**	
Ocean trenches	1.2
3. **Ocean Floor Features**	
Ocean Floor Trachite Zones (Transform faults)	—
Oceanic volcanoes	—
Continental rise	3.8

The features listed above as amongst them, the mountain ranges and volcanic island arcs (shown with asterisks in the above table) can be referred to as major landforms. In addition volcanic landforms, deltas, sandy and icy deserts and rift valleys are the other important major landforms.

The elevated mountains and plateaus of the continents formed within the last 65 million years and the volcanic island arcs delineate two belts each of which belong to a great circle round the earth. The Mediternean Asian Mountain belt follows the ALPS, the Himalayas, Indonesia, New Guinea, and New Zealand, whereas the Circum-pacific island are belt extends through the Philllippine island, Japan, Alaska, the Rocky Mountains, the Andes and Antarctica. In addition, volcanoes are associated with ocean ridge systems, and may are scattered over the ocean floors.

The continents, the mountain chains and the ocean floor features have varied relief. A hypsographic curve showing the height and depth of the earth's surface shows 29% of the total surface is above sea-level. Many processes have operated and still operating, to maintain the relief of the continents.

On a map of world's structural regions five main types of structures with distinctive landforms can be recognised which are as follows:

1. Orogenic fold belts in their first phase of development.
2. Old orogens, which suffered several phases of uplift and still from uplands.
3. Marginal fold belts on the continental side of the orogenic belt.
4. Platforms which are blocks of rigid sheild rocks with a cover of sedimentary units.
5. Shields, without a sedimentary cover.

1. OROGENIC FOLD BELTS IN THEIR FIRST PHASE OF DEVELOPMENT

When the sediments deposited in a geosynclines suffer uplift, erosion is rapid. The sediments may be folded and faulted. In the absence of lithification, it is only the contrasts between limestones, clastic rocks and igneous intrusions which are etched out by erosion, e.g. the Alps in Southern France, but in terrains like the Zagros Mountains of Iran with broad open anticlines and synclines in clastic rocks and limestones, the landforms here are in a different part of an orogen.

Thick limestones constitutes the outer units of many anticlines and the synclinal depressions are filled with molasses type of sediments. In the thrust zone to the north-east igneous and metamorphic rocks were forced to the surface. The drainage patterns are characteristics in these different terrains.

2. OLD OROGENS

The rocks of old orogens are more rigid. Old orogens are normally compoed of palaeozoic rocks which are folded, intruded and metamorphosed to schists and gneisses. Often these orogens have been planned by erosion to an even erosion surface. In the Scandinavian highlands, old fold belts are uplifted and their cover beds are ended. An orogenic belt may have an early fold and late faulting phase of development. For example, the zagros has a western fold belt of young growing folds in the Mesozoic and Cenozoic rocks; its thrust zone is in paleozoic rocks which are undergoing faulting and uplift of an earlier belt.

3. MARGINAL FOLD BELTS

Parallel to the major orogenic fold belts and on the continental side (example are, the sub-Andean fold belts of NW Argentina, lying around the margin of eastern cordillera; the foot hills east of Canadian Rocky mountains), sediments derived from the rising mountains or marine deposition maybe involved in open folding on Pre-Cambrian or Palaeozic rigid basement rocks (for example, Jura Mountains of NW Switzerland and Eastern France, marginal to European Alps).

The sub-Andean Strata have been thrust faulted and folded. Valley floors are still being filled with deposits from rising ranges. The characteristics of marginal fold belts are:

(a) Axes approximately paralled to the main mountain belts.
(b) Trellis drainage.
(c) Homoclines forming the uplands.
(d) Anticlinal strike valleys.
(e) In areas of uplift, perched synclines from upland ridges.

4. PLATFORMS

Platforms are blocks of rigid shield rocks with a sedimentary cover. These are formed due to geosynclinals belt accretion (mostly during Pre-Cambrian) with sedimentary cover occupy nearly 75% of the earth's dry land surface. The cover can be several thousands of meters thick (e.g. inland delta of Niger River). Folds in these sedimentary cover have low dips. Broad up warping giving rise to large, low angle domes or down warping to form broad basins, is characteristic of platforms. Below the sediments, the basement rocks are subject to faulting and broad warping. Draping of sediments over the faults influences the landforms. They are considered as tectonically stable regions.

5. SHIELDS

The major shield areas of the earth are constituted by the outcrops of the basement. Shields are generally brittle, rigid granitic, gneissic and associated rocks. They lack a sedimentary cover (e.g. the Beltic shield, the Indian shield, the Canadian shield). Faults and major joints become zones of selective erosion by streams. Upward or downward warping can occur over periods of millions of years giving rise to broad sags (e.g. Central Australia) and large domes, the best example is Scandinavia.

Shields are characterised by extensive Eros Ional plains. Elevations of shields may be due to up warping, block faulting, or extrusions at the rifted margins. For example, the African and Indian shield where plateau basalts, cover the ancient rocks.

6. VOLCANIC LANDFORMS

The outflow of lava takes place through the volcanic systems. According to their structure, two types of systems are recognised, viz. the *central type* and the *fissure type*. In the first type conical mountains are common, and eruptions occur through central outlets; in some large volcanoes the main vent has sub-ordinate ramifications called **parasitic volcanoes.** When tops of cones are blasted off or vast depression are formed by sinking, calderas are formed. They can be up to 15 km in diameter.

Fissure type volcanoes lack the central channel and the lavas come out at the surface through deep and extended fissures in the earth's crust giving rise to thick lava sleets. The plateau basalts of the Deccan region of India belong to this type.

There are seven types of volcanoes are recognised which are based on the character of explosions.

1. *The hawaiian and strombolian type*: The lava is of basic (basaltic) composition with temperatures of 1200° C; the lava forms

streams often up to 40–50 km long and is poor is gases and explosions are almost absent.

2. *The Etna–Vesuvius type*: The lavas are of the intermediate (andesite) composition and rarely of basic composition; they are viscous, slow in their movement and the temperature is about 1000° C often they block the vent and powerful explosions take place, ejecting large quantities of solid products like ash, lapilli, bombs, etc. The lava forms tongues on solidification.

3. *The Pelean type*: These lavas are andesitic in composition, highly viscous and temperatures are about 800° C. The process of solidification takes place in the crater also. The accumulated gases drive the lava and big gaseous explains take place ejecting large quantities of solid products and very hot gases.

4. *The Krakatoa type*: The acid and highly viscous lava does not flow out, but is ejected through big explosions as different solid products. The explosions destroy the cones and form calderas.

5. *Volcanic pipes:* These are monogenic volcanoes as they are limited to a single explosion without any emergence of lavas. The diameter of the pipe is of the order of 80 to 100 m. They are filled with solid erupted products, e.g. kimberlite pipes bearing diamonds in South Africa.

6. *The fissure type*: The lava is basic (basaltic) liquid and highly mobile – flowing out through fissures, in different directions. The lava fills all topographic depression and forms sheets over large areas. There is no explosives activity.

7. *Postvolcanic phenomena*: When the volcanic activity ceases. Then dying stages are characterised by emission of gases, geysers, lot springs and mud volcanoes.

It is now estimated that there are 4000 extinct volcanoes and about 540 active volcanoes in the world today. They have an irregular distribution pattern. No volcanoes are present in the European part and Weston Siberian part of the CIS, whereas the pacific coasts are rich in volcanoes. Zones of recent volcanic activity are related to definite tectonic mobile parts of the crust and related to deep faults. On the continents the zones of young fold mountains continental margins and island arcs are characterised by volcanic activity.

As there are several volcanic zones are found in various parts of the globe but important of them which are recognised as follows:

(a) *The Pacific Zone*: The pacific zone is also called *pacific fiery ring*. It extends from Kamchatka peninsula, the Kurile islands, the Islands of Japan, Philippines, New Guinea, New Zealand and the Solomon islands through NE tip of Antarctica, Tierra del Fuega, the Andes and Cordillers Moutains to Alaska and Aleutian islands.

(b) *The Mediterranean and Indonesian Zone*: This extends from Alps across Apennines, Caucasus, the mountains of Asia Minor and the islands of Malay Archipelago. Most of the active volcanoes of this zone are concentrated in the eastern part—the Java, Sumatra, Lesser Sunda and Southern Moluccas islands.

(c) *The Atlantic Zone*: The Atlantic zone is associated with the mid-Atlantic ridge and the transform faults associated with it. It includes the volcanoes in Icelands, the Azores, the Canary islands, Cape Verde islands, the Antilles.

(d) *The Indian and the African Zone*: The Indians and the African zone represents the volcanoes of the islands of the Indian ocean (e.g. Comoro island, islands of Mauritius, Reunion, St. Paul island) located on the mid-oceanic ridge system of the Indian Ocean. This also includes the volacanoes of Interior Africa–Kenya and Kalimanjaro.

7. ISLAND ARCS

Portion of certain arcuate submarine rides extending above sea level from the island arcs. They are common and well developed in the western pacific and West Indies areas.

On the convex side of the arcs are located the deep sea trenches or troughs (also called fore deeps) the greatest ocean deeps. Three distinct belts are associated with island arcs.

(i) The outer belt of deep trenches or fore deeps, have negative gravity anomalies are common.

(ii) The tectonic island arcs—these have positive gravity anomalies.

(iii) The inner belt of volcanoes—these also exhibit positive gravity anomalies.

The island arcs represent young and active organic sections of the earth's crust. Studies of the gravity anomalies suggest that the negative anomalies may be attributed to the accumulation and under thrusting of sediments, the positive anomalies are attributable to the higher density of the lithosphere slab that is under thrust.

There is difference in the types of volcanism in these island arcs. The arcs of Japan, Kamchatka and Indonesia are marked by landsite volcanism, the younger arcs show basaltic and basaltic andesitic volcanism. Some of the islands in these arcs have a long history of volcanic activity. The Honshu arc of Japan lates back to Ordovician. It is considered that the islands have grown by progressive addition of volcanic material from successive arc systems. In general, islands arcs are characterised by holistic, calc-alkalis and alkalis types of volcanic rocks. Islands arc disseminated copper deposits are reported from the the Philippines, South Pacific island arcs and New Guinea. Miyashiro recognised paired metamorphic belts of similar age but different types in the island arcs, particularly around the pacific ocean–an inner low pressure belt (with andalusite and granitic rocks) and an outer high pressure belt with aucophane, amphibolites and surpentinites.

8. DELTAS

Deltas are prominent areas near the margins of continents and sometimes extend far into the continental shelf, as in the case of the Ganga delta. A preliminary classification based on shape recognise five types of deltas which are as follow:

(i) *The Arcuate Delta*: The word delta is derived from the shape of the Greek letter Δ. An actuate delta has a three sided outline with the seaward margin convex or actuate. The Nile and the Niger deltas belong to the actuate delta.

(ii) *Estuarine Delta*: Estuarine delta is funnel shaped, has a length several times greater than its maximum width. For example, the Mackenzie river delta. NW Canada. The distributaries are braided or anastomosing and are separated by sand bars or island, which locally may be salty. Estuarine sediments may accumulate in certain river mouths, where the directions of stream flow is reversed by daily tidal incursions Estuarine deltas are formed when the opposing forces of stream and tidal energies are nearly equal and where the streams transport sand and silt size sediments, in addition to day size material.

(iii) *The Bird's-foot Delta*: Delta, which has the shape of a bird's-foot. The distributaries are relatively a few in number and the width of the individual channel is great as compared to its depth. Such deltas are built into deeper waters than in the case of the actuate type. The Mississippi delta is a typical example of this type as in Fig. 8.1.

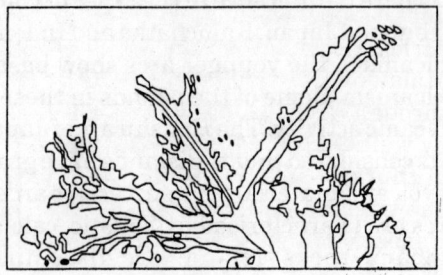

Fig. 8.1: *Bird's foot delta.*

(iv) *The Lobate Delta*: This type of delta covers river derived sediments, which have an attenuated distribution in a basin-ward direction, e.g. ancient Mississippi river delta during pleistocene. During the stages of lower sea level, the ancient Mississippi river eroded a canyon across the Subaerially exposed shelf of the Gulf of Mexico. The pronounced bulge of the water depth contours indicates the probable presence of this submarine delta.

(v) *The Cuspate Delta*: The cuspate delta has triangle shaped deposits accumulate on either side of the main channel as it accretes seaward from the original shoreline, it is split down the middle by a single channel without any significant distributaries, e.g. the Brazos river delta, Texas, the Tiber river delta of Italy.

9. THE SANDY AND THE ICY DESERTS

The sandy and the icy deserts of the globe constitute significant types of landforms. The immence desert belt of the world extends from the shore of the Atlantic in a giant Scimitar Curve across north of Africa, Arabia, Turkestan and to the Gobi. To the north of Central Eastern Europe into Siberia and merges into the close packed ice-floes of the Arctic.

The North Pole, with monotonous broken ice merges into its surrounding landmasses. Round the Arctic lies the expense of frozen north lands of Russia, Greenland and Canada. Seen from the South Pole, the Southern Hemisphere is dominated by the Southern ice cap which covers an island larger than the continent of Europe.

These sandy and icy deserts of the world show some characteristics landforms which can be easily recognised.

10. RIFT VALLEYS

Rift valleys are very large structurally controlled landforms formed due to down-faulting, with fault scraps on either flank. The East Africa rift valley system is about 3000 km long extending from Malawi in the South bifurcating into two arms and then extends through Ethiopia to the Red Sea. This rift valley system is marked by well-defined escarpments 400 to 2000 m high. The rift valleys of this system are 3 to 90 km wide, but the floors of the valleys show variations in depth due to transverse buckling and late Cenozoic Volcanism. The deepest parts are occupied by lakes. Some of these lakes are Saline as they do not have drainage outlets. Lake Tanganyika is 1400 m deep and its floor is 650 m below sea level.

The dead sea rift is about 600 km long and 10–20 km wide. The Rhine graben is associated with heat flow, active faults and earthquakes. Large scale rifting is the median rift of plate divergence as we see in the case of the mid-Atlantic ridge and doming is associated with upwelling of mantle material and volcanism.

9 Geomorphic Features of Indian Subcontinent

Physiographically Indian subcontinent can be divided into three well marked regions, which are as follows:

1. The Peninsula or Peninsular shield lying to the south of plains of the Indus and Ganges river system.
2. The Indo-Gangatic alluvial plains stretching from Assam in the east through Bengal, Bihar, Uttar Pradesh to Punjab and send on the west.
3. The extra Peninsula the mountainous region of the mighty Himalayan ranges and their extensions into Baluchistan in the west and Burma and Arakan in the east.

These regions can be categorised into geomorphic divisions, subdivisions and provinces depending on their characteristics geomorphic fetures, as was done by the National Atlas Organisation in 1964. This is because as these regions exhibit marked contrast in geomorphic features, stratigraphy and structure.

I. Northern Mountains

1. *Western Himalaya*

(a) North Kashmir Himalaya	: High relief; Valley glaciers; snow fields, intermont basins.
(b) South Kashmir Himalaya	: High relief; local alpine peaks; long narrow valleys and lakes.
(c) Punjab Himalaya	: High relief; accordant summits; long narrow valleys; gorges, river terraces.
(d) Kumaon Himalaya	: High relief; accordant summits; local alpine peaks.

2. *Central Himalaya*

   (a) Nepal Himalaya (West) : High relief; alpine peaks.

   (b) Nepal Himalaya (East) : High relief; alpine peaks (some of the highest peaks in the world); piedmont belt to the earth.

3. *Eastern Himalaya*

   (a) Eastern Himalaya (West): Strong to high relief; piedment belt to the south.

   (b) Eastern Himalaya (East): Strong to high relief; gorges, terraces.

4. *North Eastern Range*

   (a) Purvanchal : Folded strong and weak strata; strong relief; foot hill plains.

   (b) Meghalaya : Uplifted and dissected plateau; moderate to strong relief.

   (c) Assam valley : Flood plain; alluvial terraces.

II. Great Plains

1. *Western Plains*

   (a) Marusthali : Sandy desert; shifting sand dunes.

   (b) Rajasthan bagar : Desert alluvial slopes; hills; moderate relief; pediments' palayas.

2. *Northern Plains*

   (a) Punjab plains : Extensive plains with meandering rivers.

   (b) Ganga, Yamuna Doab : Extensive plains, flood plains.

   (c) Rohilkhand plains : Plains with dry river beds in the northern part.

   (d) Avadh plains : Extensive plains with meandering rivers.

3. *Eastern Plains*

   (a) North Bihar plains : Plains with ever changing river courses.

   (b) South Bihar plains : Plains a few isolated hills in the Southern part; low relief.

(c) Bengal basin	:	Minor river terraces; young marine plain with tidal estuaries.
(d) North Bengal plains	:	Piedmont alluvial fans and dissected plains; low relief.

III. Central Highlands

1. *North Central Highlands*

(a) Aravali range	:	Polycyclic mountains on folded strong and weak strata; moderate to strong relief.
(b) East Rajasthan uplands	:	Peneplains with low monadnocks.
(c) Madhya Bharat pathar	:	Rolling plateau with young incised valleys.
(d) Bundelkhand upland	:	Eroded upland of crystalline rocks low to moderate relief.

2. *South Central Highlands*

(a) Malwa plateau	:	Lava Capped plateau and buttres; terraced slopes; moderate relief.
(b) Vindhayan range	:	Old sandstone plateau; young lava plateau; moderate relief.
(c) Narmda valley	:	Rocky narrow canyons and gorges; alluvial basins.
(d) Vindhya scarplands	:	Even created ridges and long valley belts; moderate relief.

IV. Peninsular Plateaus

1. *North Deccan*

(a) Satpura range	:	Tilted lava flows; fault blocks; strong relief.
(b) Maharashtra plateau	:	Residul plateau; moderate to strong relief on linear range of crystalline rocks.

2. *South Deccan*

(a) Karnataka plateau	:	Residual plateaus; extensive peneplain on massive crystalline rocks.
(b) Telangana plateau	:	Residual plateaus; extensive peneplain on massive crystalline rocks; a few monadnock.

3. *Eastern Plateaus*

    (a) Bhaghelkhand plateaus : Isolated flat topped hills over maturely dissected plateau; moderate relief.

    (b) Chotanagpur plateau : Extensive plateaus at different levels; deeply dissected; strong relief; rolling plains to the south-east.

    (c) Mahanadi basin : Large basin flanked by horizontal to low dipping strata; moderate relief.

    (d) Garhjat hills : Repeatedly uplifted and discussed peneplain on strong rocks; strong relief; trenched by rivers.

    (e) Dandakaranya : Submaturely dissected plateau of strong and weak strata; strong relief.

4. *Western Hills*

    (a) North Sahyadri : Flat summit; terraced surfaces; lava flows finger like spurs; strong relief.

    (b) Central Sahyadri : Accordant summits on dissected rocks of complex structure; moderate to strong relief.

    (c) Nilgiri : Rolling surface on complex crystalline rocks with strong relief; young incised valleys.

    (d) South Sahyadri : Mountains over folded strong strata; longitudinal and transverse ranges; strong relief.

5. *Eastern Hills*

    (a) Eastern Ghats (North) : Peneplained and uplifted crystalline rocks of complex structure; strong relief.

    (b) Eastern Ghats (South) : Submaturely dissected horizontal to folded rocks in the north with moderate to strong relief; peneplianed and uplifted plateaus of strong relief in the south.

(c) Tamil Nadu uplands	:	Bevelled and maturely dissected hills with moderate relief; isolated hills; pediments.

V. West Coast

1. *West Coastal Plains*

(a) Kutch peninsula	:	Submaturely dissected coastal plain; partly submerged to the north; hills of moderate relief in the interior.
(b) Kathiawar peninsula	:	Young to mature belted coastal plain with hills of strong relief in the interior.
(c) Gujarat plains	:	Alluvial plains dotted with isolated hills; young marine plain.
(d) Konkan coast	:	Rocky broken coastal belt with submerged border and rocky islands; moderate relief.
(e) Karnataka coast	:	Dissected and traced coastal plain partly submerged; low to moderate relief.
(f) Kerala plain	:	Young to mature belted coastal plain, partly submerged; low relief.

2. *Western Continental Shelf*

Broad submarine platforms.

VI. East Coast

1. *East Coast Plains*

(a) Utkal plain	:	Terraced coastal plains and sand dunes; alluvial plain.
(b) Andhra plain	:	Young coastal plains with food plains and deltas.
(c) Tamil Nadu plain	:	Alluvial plain uplifted and dissected, young marine plain.

2. *Eastern Continental Shelf*

Narrow shelf with submarine canyons.

VII. Islands

1. *Arabian Sea Islands* : Coral islands.

2. *Bay Islands*   :   Elevated portions of submarine
                                     mountains, a few volcanic cones.

In general, geomorphic features are categorised in three degrees of relief which are as follows.

   1. *High Relief*   :   High relief is measured in
                                       thousands of feet.

   2. *Moderate Relief*   :   Moderate relief is measured in
                                       hundreds of feet.

   3. *Strong Relief*   :   Strong relief may be anything
                                       approaching 1000 feet with a wide
                                       latitude on both sides.

10 Geomorphology and the Construction Industry

Geomorphology deals with the evolution, description and interpretation of earth's topographic features which play an important role in planning and completing engineering projects. The understanding of topography is a powerful tool for the success of many construction programmes.

Many site selections are influenced by topography which gives the clues to locate construction materials and indicates hazardous condition whose recognition allows rejection of certain sites for specific engineering projects present day geomorphic studies are mainly includes interpretation topographic forms and studies mainly in pedology, meteorology, geography, engineering and related disciplines.

PRELIMINARY GEOMORPHOLOGICAL STUDIES

On the basis of geomorphological study of topographical maps and aerial photographs establishes the nature of the topography, materials likely to be available as bad rock and surfacial, geological structure, surface and subsurface hydrology, etc. During this preliminary stage, sites unsuitable for engineering projects are rejected, thus bringing about saving in time and money; project costs can be tentatively estimated and construction designs and operations are generally be evaluated.

DETAILED GEOMORPHOLOGICAL STUDIES

Detailed geomorphological studies provide date when final site is selected. Different aspects of data which are as under.

1. Nature and availability of construction materials.
2. Probable geological hazards.
3. Conditions of the foundation.
4. Localised site location studies.
5. Inter-relation of sites and their environment.

Nature and Availability of Construction Materials

When we understand the active geomorphological proneness, the geomorphological stage of the area is determined and the factors controlling the topographic development are evaluated. It is found that topography often throws light on the nature of the surface material on which geomorphic agents operate. The degree of weathering and erosion, slope characteristics, drainage patterns, relief differential help in identifying the materials.

Alluvial fans, pediments, shore lines, lake bottoms, knobs and kettle moraines, kames, eskers, karst and abandoned stream channels suggest the possible type of surface and subsurface materials that will be available. Topographic highs indicate rocks that are resultant to weathering and erosion, suggesting potential construction material. Gravel often occurs as a capping on isolated butte-like features. Limestone, dolomites, sandstones and lave flows are common steep escarpments.

Trellis drainage pattern is seen in areas of folder and eroded sedimentary rocks; dendritic patterns are characteristic of areas with homogeneous rocks. Carbonates often show drainages controlled by joints.

The porosity and permeability of surface and subsurface materials are indicated in the drainage texture. Fine-texture drainage is seen on non-porous, impermeable claystones which may produce bad land topography. Coarse-textured drainage suggests a porous and permeable lithology, e.g. gravel. Sandstones with intermediate porosities and permeabilities develop median-textured drainage.

The nature of slopes, the slope angles especially, frequently suggest the histology. Easily weathered materials (e.g. shales) suggest low slope angles. Materials that resist weathering and erosion develop steep slope angles. In massive materials, weathering gives rise to large rock blocks removed from outcrop surfaces, whose slope angles tend to be steep or vertical. The most satisfactory approach for the use of topography as an indicator of lithology involves recognition of specific landforms, drainage characteristics and slope angles.

Probable Geological Hazards

The recognition of geological hazards is an important application of geomorphpology to construction geology. These mainly include criteria for the recognition of the following.

Slope failure: Previous history of slope instability and materials amenable to slope failure.

In hummocky, topography, stable slope are generally uniform and reflect relatively simple slope angles. Unstable slopes are often complex.

Slope failure is associated with hydrological activity and in competent materials, which can be recognised in the topography.

Before constructions, we generally estimate the runoff potential, to make proper designs for accommodating dangerous flow. Subsurface drainage problems can be recognised from topographies that include marshes, swamp and high water tables. Glacial deposits with high clay content are likely to present construction difficulties. Karst topography is notorious for drainage complexities.

An understanding of the nature and evolution of topography and rock types involved is a good approach for predicting potential construction difficulties.

Conditions of the Foundation

The main foundation problems which can be anticipated from geomorphological studies are swelling clays, drainage problems and differential compaction. The nature and thickness of surface or near surface material can be predicted by understanding the origin of topography. Recognising whether topography is due to erosion or deposition is an important aspect. In depositional topography, the suitability of materials for foundations or excavations depends on the origin of the topography.

Localised Site Location Studies

Geomorphological interpretation with the help of topographical maps, aerial photographs combined with field checks helps in selecting locations for roads, canals, dams, airports, bridges, tunnels, etc. such an interpretation indicates areas of relief, slope to be encountered and direction of alignments. We make estimates regarding required cut and fill material and access to airport sites, etc. Probability of fog in topographically low areas or indications of high wind rule out an airport site.

Inter-relation of Sites and their Environment

In project planning, geomorphological studies play an important role which is represented by the effect of engineering projects on environment. The major problem is the probable adverse effect of projects on environment, viz. flooding or slope failure or pollution of streams and atmosphere.

We pay emphasis on quaternary topography, stratigraphy and structure, to provide a good geomorphological analysis. Topography is a constantly evolving feature of the earth's surface and engineering projects, and should be compatible with its evolution.

Geomorphic Processes and Plate Tectonics

William Moris Davis (1850–1934) and Walther Penck gave primary importance to erosion as a means to evolution of the landscapes of continents.

In fact, modern geomorphic studies understandably take into account the role of the internal dynamism of the earth in the evolution of landscape features. It is considered that on the basis of the known rates of erosion of the earth's crust by the major rivers of the world, the crust would have been eroded to the base level not just once, but many times over; and the very fact that several landform features stand out boldly proves that endogenetic forces from deep within the earth have a crucial role in the evolution of landscapes. The emergence of the plate tectonics concept gave impetus to the study of landscape evolution influenced by global tectonic processes as has been highlighted by C.D. Ollier in his study work tectonics and landforms in 1981.

It is fact that the plate tectonics concept in to universally accepted by geoscientists. Nevertheless, many consider the concept as an integration of ideas like continental drift and sea-floor spreading and identify in the concept a from work with in which seemingly unrelated features can be studied with advantage.

Some of the important points of interests from the point of view of geomorphologists which are as follows.

1. Earthquakes and volcanoes are geomorphology hazard which have been very much facilitated by the practical application of the plate tectonics concept. Prediction and ever controlling the hazards have become possible. Plate tectonics concept has increasingly helped in the evaluation of resources both

on land and in the ocean bottom. Convection currents emanating from the mantle source are identified to carry an array of minerals of economic importance (Fig. 11.1).

2. Several geomorphological features are studied in relation to folding, faulting, apostasy and the like. The motive force for such tectonic features is provided by the plate tectonics concept (Fig. 11.2).

3. The plate tectonics concept offers an integrated and comprehensive explanation for the relief features on the continents and oceans basins. Features like block mountains, fold mountains, rift valleys, island arcs and so on are covered by the concept.

4. In the early evolutionary stages of geomorphology as a science, studies were mostly restricted to continental features. Now, it is rightly held that a holistic approached to the problem, encompassing both landform and ocean bottom features is necessary and the plate tectonics concept is a unifying theory linking the continent and the ocean bottom.

5. Geomorphology is a border science on both geology and geography. The plate tectonics concept has implications in the study of geomorphology. This compels both geology and geography to come closer. And when these two come closer, such closeness, especially in the context of increasing recognition of the importance of multi-disciplinary approach to problems in earth sciences is conducive for progress.

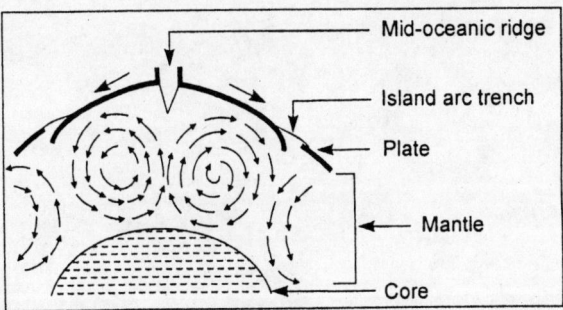

Fig. 11.1: *Convection system in the mantle causing motion of the overlying lithospheric plates.*

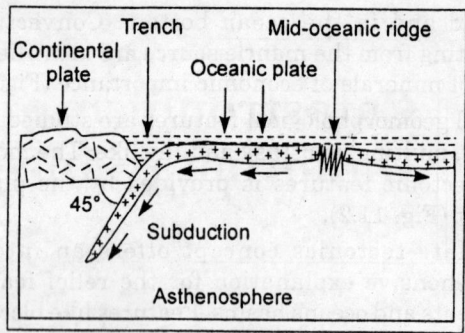

Fig. 11.2: *Plate motion, occeanic plate glides over the soft asthenosphere from mid-oceanic ridge to the zones of subduction.*

12 Catastrophism and Uniformitarianism

William Morris Davis, considered as the father of modern geomorphology synthesised the scattered elements of the subject and gave shape to geomorphology. In the Pre-Davisian era, though there was no science distinctly recognised as geomorphology, certain developments are recognised as seminal thoughts in geomorphology. Aristotle, the Greek philosopher of yore thought that some streams were the outcome of the downward percolation of rainwater. In the fifteenth century, Leonardo da Vinci observed that rivers excavate valleys by eroding their beds. However, there was no concerted development of scientific thoughts in the middle ages when religious beliefs were interwoven with science. The earth as a cosmic body was thought to have evolved within a period of just six days of creation, and that man made his appearance on the face of the earth in the year 4004 BC. Thus, the theory of catastrophism which tersely explained the evolution of natural orders to quick and well-defined happenings came into being. The spirit of scientific enquiry whose motto is "to strive, to seek and not to yield" had not dawned. Despite progress in many scientific field, the theory of catastrophism held sway almost to the end of the eighteenth century. Then the theory of uniformitarianism gained slow ground.

According to the theory of uniformitarianism, the natural laws and processes which were operative in the geological past are essentially the same as the ones recognised to be in operation in the present time. The quintessence of the concept is "the present is the key to the past".

Applied to the science of geomorphology, it would necessarily mean that the landforms of the present day have evolved by processes operating over a long span of time. Thus, uniformitarianism is the antithesis of catastrophism. Hutton (1726-1789), a Scottish geologist, propounded the theory of uniformitarianism and Charles Lyell (1797-1875), another Scottish geologist put forth persuasive arguments in favour of the theory in his well-known book entitled, "Principles of Geology" both Hutton and Lyell have views which were based on field studies in coal fields. The latter described the processes of disintegration of rocks, transport of sediments by fluviatile action and the deposition of sediments. John playfair another exponent of the theory outlined drainage networks. "Playfair's Law" puts forth that areas which have uniform bed rock and structure, when subjected to river erosion for long give rise to valleys proportional in size to the streams in the areas and that stream junctions are concordant.

Infact old customs die hard. Even at the dawn of the nineteenth century, the origin of the Niagara gorge was attributed to a sudden catastrophe. Some were intrigued by the absence of rivers in well-defined valleys. However, persistent scientific enquiries allayed many lingering doubts and today it is recognised that landforms have resulted by inexorable action of natural agencies.

13 Working Stages of Rivers

Rain falling on the dry ground first moistens the soil before any water may be evaporated or absorbed into the earth.

But until the late seventeenth century, the continuous flowing of water from the earth to the rivers was thought to be magical. It was concluded that there was more than enough rain and other precipitation falling on the catchments area, the entire region drained by a river to account for the flow of the river.

Modern facilities in research areas brought the some unexpected facts. In 1960, R.L. Nase, an American hydrologist, calculated the amount of water in storage and in transit in various parts of the hydrological cycle, the continuous process in which water evaporates from the sea, is precipitated on to the land, and eventually runs back to the seas as shown in Fig. 13.1.

The quantity of water to be found both in and on the continents is only three percent of all the earht's stock. Of this, over 77 percent is locked up in the ice caps and glaciers and another 22.5 percent lies under the ground.

Running water is the most widespread and effective agent of landscape sculpture, continually caring deep valleys and broad plains and altering the surface features of the earth. In general, work starts with the rainfall and other forms of precipitation which provide, directly or indirectly, all the waters guided into natural channels to form streams and rivers. The source of a stream may be immediate run-off from rainfall, or subsurface water from a spring or the release of water held temporarily in lakes or ice.

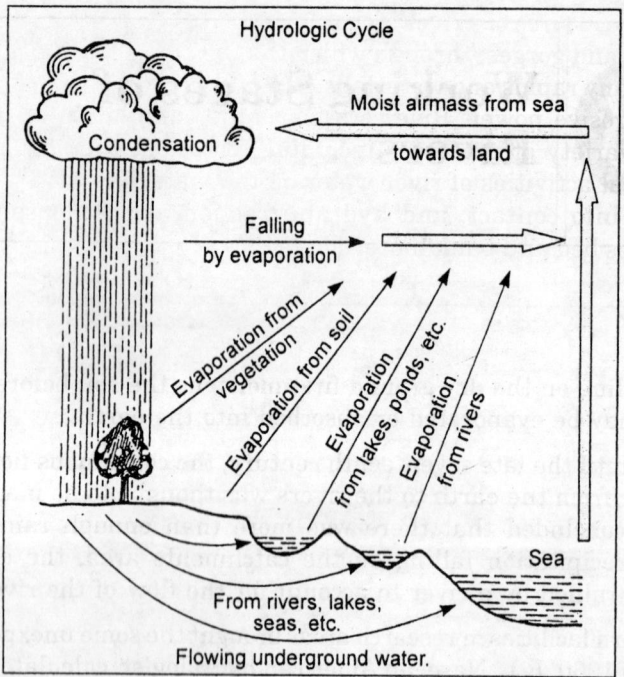

Fig. 13.1: *Hydrologic cycle. The entry of water into atmosphere by evaporation from the sea and its return from the atmosphere to the sea.*

STUDY OF THE STRUCTURE OF A RIVER BY SEPARATION INTO PARTS

Rivers develop a distinctive slope or gradient on their course to the sea. Throughout a river's course, material is constantly being eroded from some sections and deposited in others. Where the river beds are very irregular and full of holes. It is likely that erosion is taking place, where the beds are flooded with alluvial material, deposition is taking place. Erosion occurs where the streams has an excess of energy. When its energy is decreased, by a fall in gradient or by an obstacle, such as the still water of a lake, the stream is no longer competent to transport its debris and must start to deposit it.

As we know that a river is often described as passed through successive stages of development, from youth to maturity and on the old age. Because of the erosive energy of youthful streams, the upper course of a river is characterised by steep sided, V shaped valleys and gorges, through which the water rushes and tumbles over many rapids and waterfalls. Running water by itself has very little erosive power. River erosion is the cumulative effect of a great variety of processes including corrosion, the solvent and chemical activities of river water on the materials with which it comes into contact, and hydraulic action, which loosens and removes bed and bank material.

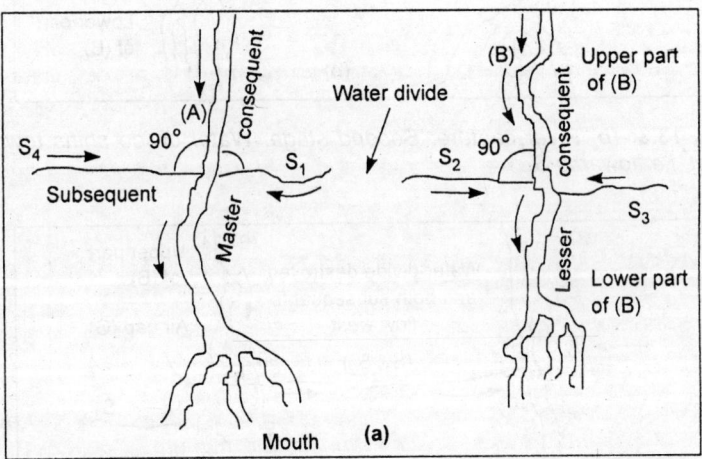

Fig. 13.2: *River capture: First stage. (a) See the conditions of the following streams: (i) Master consequent (A), (ii) Lesser consequent (B), (iii) Subsequents (S₁, S₂, S₃ and S₄).*

Once armed with loosed debris, running water becomes very destructive. Corrasion (not to be confused with corrosion) is the process by which the river bed is worn away by the boulders, pebbles, sands and silt being carried along in the stream, Attrition is a mechanism in river erosion, whereby the transported materials or load are themselves broken down into smaller pieces by the continuous bathering they receive as they move downstream.

A river's load can be transported in four different ways, by traction where the load is rolled along the stream bed, by saltation where the particles are moved in a series of short jump, by being held up in suspension, and in solution, dissolved in the water. As the velocity of the river slackens, the larger boulders and pebbles

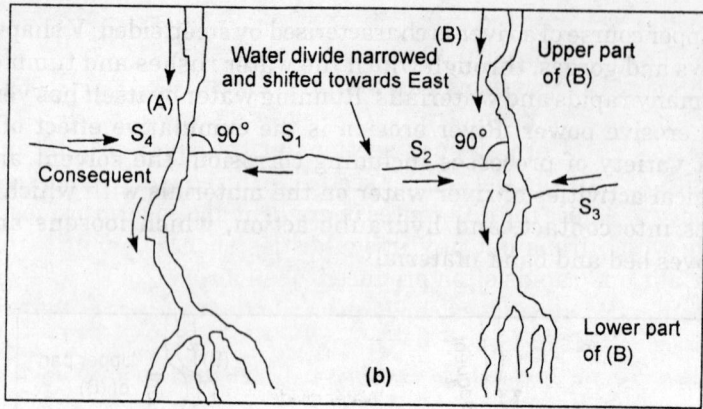

Fig. 13.3: *(b) River capture: Second stage. Water divide shifts towards east, i.e. towards S_2.*

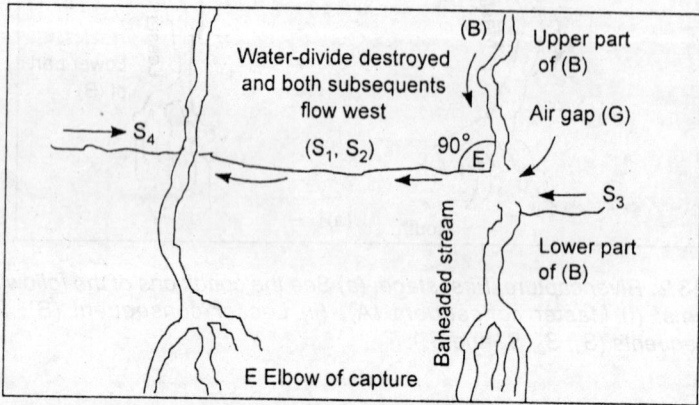

Fig. 13.4: *River capture. Subsequent S_2 reverses its direction and joins S_1. Consequent (B) is captured. Air gap (G) is formed. See the beheaded stream. Also note the elbow of capture (E).*

come to rest, leaving only the material in suspension and solution to continue its journey downstream.

The nature of the bed and land adjacent of a natural river may indicate erosion or deposition although the river itself may appear to be incapable of either. But if the same river is looked at in flood, the large and apparently immovable boulders are being bounced along the bed, and material is being deposited in the lower reaches. In this way an estimated 8,000 million tones of eroded rock waste are transported from all parts of the world to the sea every year. This represent an annual loss corresponding to 77 tonnes per square kilometre.

Erosion cuts into steeper sections of the river profile and deposition occurs on the gentler sections, consequently, there is a tendency to reduce the differences in gradient and so form a smooth profile which irons out any irregularities such as small lakes, waterfalls and rapids often produced when a river course crosses resistant rock bands. The deepening of the river valleys by erosive processes is limited by sea level, which acts as a lowest point or base level for erosion. All rivers attempts to smooth out their long profiles from source to mouth. When a river attains this condition it is said to be at a state of equilibrium or grade and the profile is known as a graded profile.

RUNNING WATER IN MATURITY STAGE

When running water reach middle age or maturity, sideways or lateral erosion becomes more important, broad valleys with gentle slopes are produced and the rivers tend to meander. A meander is the name given when a river follows the natural gradient of the land and forms semi-circular bends with geometrically perfect curves.

In past days, there was a thought that chance irregularities in a river's curve where the cause of meanders always confirm to a pattern. Laboratory experiments have shown that even those rivers developing on a uniform slope of uniform sediment are always found to modify their channels by erosion and deposition so that a series of symmetrical bends develops.

Braiding is another feature of a mature river valley. This occurs when a river which is laden with debris emerges from a ravine onto a bordering plain. The velocity of the river is suddenly checked by the abrupt change of gradient and much of its sediment load is dropped. The large volume of debris thus deposited obstructs the flow of the river which divides into distribution which continually separate and unite.

Meandering rivers develop their flood plains by depositing salt and alluvium as they migrate sea-wards. Further deposits known as overbank deposits are left on the flood plain by flood waters each time the river overflows its banks. The flow of the water which floods onto the banks is not controlled by the main current

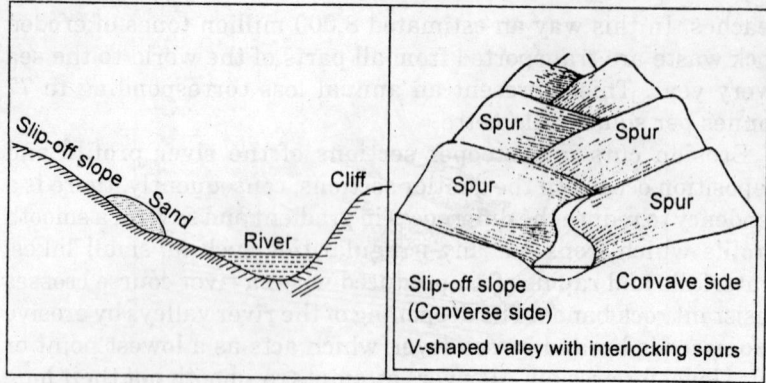

Fig. 13.5: *Cross-section of river. Deposits of sand, gravels, etc. on the convex sides.*

Fig. 13.6: *Interlocking spurs.*

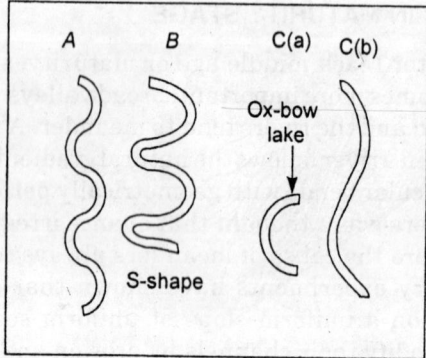

Fig. 13.7: *The development of Ox-bow lake. A meander, B the S-shape of meander, C (a) Oxbow lake, C (b) the course of the stream is straightened.*

since it is outside the river channel. As the flood slows down, it drops the coarsest part of its load, gradually building up a low embankment or levee on either side of the river. As the levee develops so too does the river bed. Silt, carried down by flood waters too small to overtop the levees, is deposited on the river bed as the flood subsides. If, however, the flood waters do breech the levees the consequences can be disastrous.

Running water are rarely allowed to achieve a state of equilibrium or grade. More often they are interrupted by changes

Fig. 13.8: *Alluvial cone.*

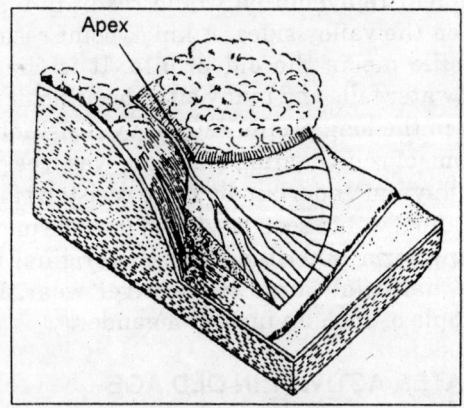

Fig. 13.9: *Alluvial fan.*

of base level, caused either by a change in the sea level or by earth movements which have uplifted or lowered the earth's surface. If a region is depressed by earth movements, its surface is brought nearer to base level, and because the work to be done by erosion is diminished, the stages of the cycle then in progress are passed through more quickly.

On the other hand, if a river that has already established a flood plain is rejuvenated that is, the base level is lowered the river will cut though the alluvium onto the rocks below in an attempt to regrade its course. The margins of the original valley floor are then left as terraces above the level of the rejuvenated

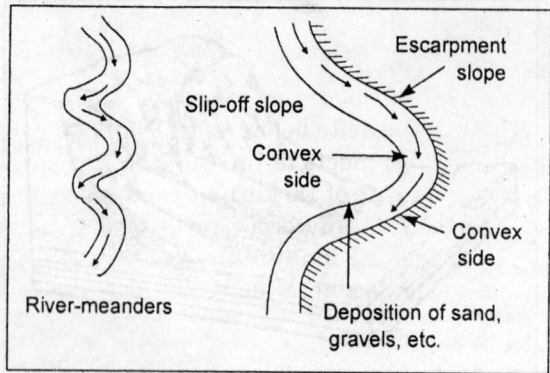

Fig. 13.10: *River-meanders, escarpment slope and slip-off slopes.*

river. Subsequent rejuvenation would result in a second pair of terraces left on the valley sides. A knick point results where the regarded profile meets the old profile. It is characterised by cascades and water falls and many often be difficult to distinguish from a break in the long profile caused by resistant rock bands.

If at the time of rejuvenation a stream was freely meandering on a floor of alluvium underlain by more resistant formations, the deepening channel would soon be etched into the underlying rocks with a winding form inherited from the original meanders. In England the "hair pin" gorge of the river wear, Durham, is a familiar example of such an incised meander.

RUNNING WATER ACTIVITY IN OLD AGE

In old age, the wide, heavily-laden river glides sluggishly over a very broad, almost level plain. Its main work is deposition. Most of the sediment load of a river is carried out to sea or to a lake where the velocity is checked and provided that deposition is at a greater rate than removal by currents, much of the load is deposited an alluvial tract called a delta.

There are three basic types of delta formation.

1. If the river water is denser than the sea or lake because of its load of sediment, it flows along the bottom and forms in elongated delta.

2. If the river water has the same density as the sea or lake, it spreads out in the shape of a fan and an arcuate delta, like that of the river Nile is formed.

3. Finally, if the river water is less dense, it makes a few confined channels for itself and a bird's foot delta, like that of the Mississippi, is formed. Deltas can grow with great rapidity. For example, the delta below Astrakhan in the Soviet Union, where the Volga meets in the Caspion sea, was at one time growing at the rate of 1.6 km every five or six years.

Deposition does not only occur in the lower parts of a river course. Many youthful mountain rivers descends steeply to neighbouring lowlands where they drop their suspended load to form alluvial fans or steepsided alluvial cones. Where closely spaced streams discharge from a mountain region their deposits may eventually unite to form a continuous plain or piedmount alluvial plain. For example, the Indo-Gangetic plain in India extends from the delta of the Indus to that of the Ganges–Brahamputra as shown in Figs 13.11, 13.12 and 13.13.

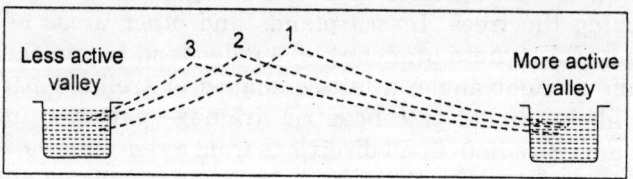

Fig. 13.11: *Shifting of water-divide.*

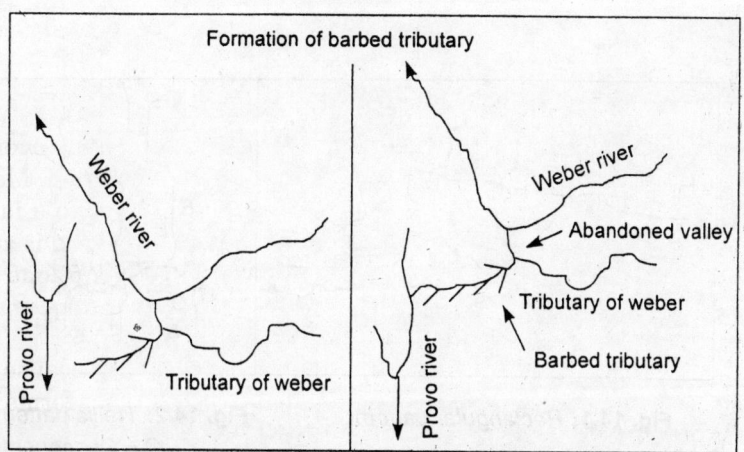

Fig. 13.12: *The shifting of river provo (more active) towards the tributaries of the weber.*

Fig. 13.13: *The provo has captured the tributaries of the weber and formed barbed wire tributaries.*

Patterns and Texture of Drainage

In areas of uniform layers of sedimentary rocks, such as clays, drainage patterns emerge which resemble the branching of a tree. The dendrite pattern (from the Greek word "dendron" meaning a tree) resemble the branching of tree resemble the branching the trees. In scarplands and other areas of gently dipping rocks where branches of smaller tend to join the larger streams at right angle, a lattice shaped or trellised pattern is more likely. Radial or concentric drainage patterns, in which the channels radiate in all directions from a central area, develop on the slopes of volcanoes or conical-sheped hills as shown in Figs 14.1 and 14.2.

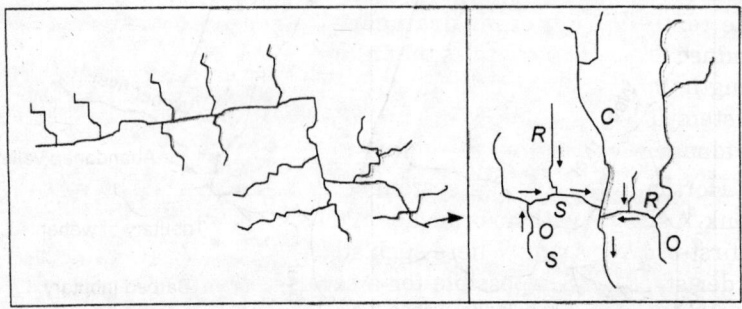

Fig. 14.1: *Rectangular pattern.*

Fig. 14.2: *Trellis pattern.*
O = Obsequent
R = Resequent
C = Consequent.

The relative spacing between the streams of river's network is aptly described as drainage texture. Fine textured drainage indicates a closer network of stream channel than coarse-textured drainage. Drainage texture is influenced by many factors, including climate. For example, area which receive all their rain in short, sharp thunder-storms often develop a fine textured network. Texture is also affected by rock structure. Drainage lines are more numerous over highly impermeable surfaces such as clays which do not allow water to soak into them than over permeable rock-like chalk or limestone as shown in Fig. 14.3.

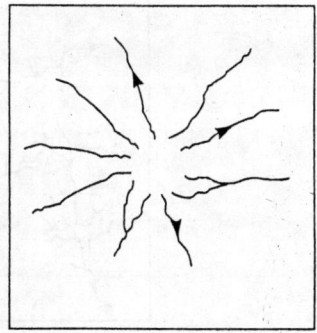

Fig. 14.3: *Radial pattern.*

Radial Pattern

The bad lands topography in parts of the western United States illustrated one set of conditions which leads to the development of fine texture—impermeable clays, sparce vegetation and rainfalling in violent thunderstorms. Coarse drainage texture is well displayed on the sand and gravel outwash plains in front of glaciers.

In 1945, a useful method of measuring drainage texture which was suggested by R.E. Horton. He developed several numerical measures so that objective comparisons could be made between the texture of different draingage networks. For example, he defined drainage density as the value of the sum total of the stream lengths in the system divided by the total area drained by the system under investigation. In this way the texture of several drainage systems may be compared as in Fig. 14.4.

Horton also devised a system of listing streams in an order of rank. A stream with no branches of tributaries joining it was called a first-order stream. Where such streams join they form a second order stream. It is possible for a second-order stream to receive another first order tributaries without being promoted to the third order, but where two second-order streams join up then a third order stream begins, and so on. The complexity of different drainage systems can be measured by determining which have the highest order streams.

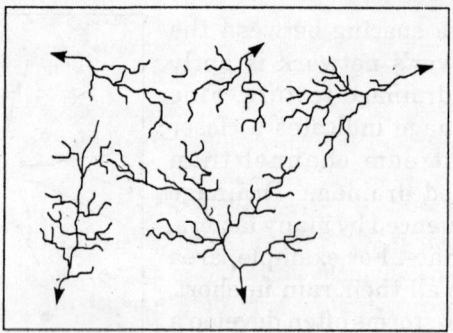

Fig. 14.4: *Dendritic pattern.*

THE ORIGIN OF DRAINAGE SYSTEM

A newly formed land surface is immediately exposed to weathering and erosion. In regions of sufficient rainfall, streams, are created by the run-off of rain. Thus, the streams and rivers whose original down hill course was determined by the initial slopes of the land are termed consequent streams, because their flow is a consequence of the pre-determined slopes. The final pattern of drainage to emerge then depends on the nature and structure of the rocks over which the consequent stream flows. If the rocks are uniformly resistant, a dendritic pattern will form the branching drainage system developing from consequent stream is known as insequent or lateral consequent.

However, it is more common that an area consists of alternating strate of hard and soft rocks. As the consequent stream cuts its way down to the sea, it behaves in a different way as it passes over differing rock beds. Narrow valleys will be carved through the more resistant rocks, which will stand up as ridges, and the stream will stand up as ridges, and the stream will wear away the clays and softer rocks to produce broad valleys.

In this situation, the tributaries feeding the main consequent stream are much more likely to take advantage of the weaker bands and to make rapid upstream or headward erosion along these. These tributories are then known as subsequent streams. A typical trellised drainage pattern then develops, with the consequent rivers flowing parallel to the gentle dip slope of the rock and the subsequent tributaries flowing at right angles to this.

The valley of a subsequent stream steadily widen between the ridges formed by the bands of more resistant rock on either side. As the valley is broadends, the gentler slope will become steeper or gentler to approximate the angle of dip of the underlying hard rock. Gradually, run off from these slopes to meet the subsequent.

These new streams are sometimes called resequent streams since they appear to be consequent in direction, but they are of a later generation, formerly only when the original incline down which the consequents flow is broken up into dip slopes. Shorter streams will also flow in the opposite direction down the steeper escarpment slope of resistant rock to join the subsequent stream, and are known as anti-dip or obsequent streams.

Inverted Relief

The land's original surface on which some drainage system develop takes the form of a series of syncline (dips) and anticlines (rises). If, for some reason, the erosive powers of river wear away the anticlines at a gtreater rate than the synclines, then the synclines eventually form the mountain and the anticline will form the valleys. This phenomenon, where the topography is opposite to the structure, is known as inverted relief, although it is in fact the more usual case in areas of folded rocks. For example, snowdown, the highest mountain in wales, has developed from a syncline structure.

During the first stage in the development of inverted relief, the main consequent stream follows the natural hollow of the syncline. However, the peak of the anticline is often so weakened or even cracked by folding that the hard bed capping the structure is easily breached near the crest of the anticline subsequent streams will easily developed here, whilst obsequent streams will begin to flow down the inward facing scarps. If the arrangement of the hard and soft beds on the anticline is suitable, erosion of the anticline by the subsequent streams may be much faster than erosion by the consequent streams in the synclinal valley. The relief now becomes inverted.

After erosion has continued for a period of time, the subsequent streams may cut down as for as a lower hard bed of rocks. Its downward path is then checked, but the original consequents may erode relatively easily through the remains of the upper hard bed and underlying soft beds. Ultimately, a pattern very like the original develops, but as the streams are not direct descendants of the original consequents, they are more properly described as secondary consequents.

Watersheds

The ridge of high land separating two neighbouring streams or river systems is known as a devide or watershed. As early as 1877, the American geologist G.K. Gilbert discovered that, in the later stages of the fluvial cycle. Water sheds between streams of similar size are uniformly spaced and stream gradients are roughly equal on opposite sides of the watersheds.

However, Gilbert found that this was not true in the earlier stages of the cycle. A stream flowing down the steeper slope on one side of an unequally inclined ridge erodes its valley more rapidly than one flowing down the other, more gentle slope. As a result the watershed gradually recedes or migrates towards the side with the gentler slope. The deviding ridge quickly takes on a characteristic ruggedness, varying in height. This concept came to be known as the law of unequal slope, sometimes the crest is lowered until it develops into a depression or cole in the devide which often becomes useful as a pass for roads built across mountain ranges.

River Capture

A strange phenomenon which has come to light in the study of drainage systems is that of river capture. This interesting example of natural piracy takes place when a major consequent river acquires a vigorous subsequent tributary which is etching its way by headward erosion back along a very weak outcrop of rock wearing back its watershed as its goes. Eventually, the subsequent stream may breakthrough the devide and intercept an adjacent river system whose upper tributaries or headwaters are captured and diverted along the course of the subsequent stream. The beheaded stream is left as a misfit since the reduced volume of water is no longer appropriate to the valley through whcih it flows. If river capture is predominent in a region, a pattern commonly occurs in which tributaries join the main river in backward looping or "boathook" bends, providing a barbed drainage system.

Most river capture is largely explained by differences in rock resistances. However, in areas of permeable rocks there is the possibility of underground diversion preceding and aiding surface capture. Active erosion at the head of a lower but vigorous stream may cause it to approach close to a higher level valley and gradually sap more and more of the upper stream's ground water supply. Tower creek in yellowstone National Park, USA, is an interesting example of the undergound diversion of drainage.

A river's course can also be diverted or deranged by the imposition of a barrier in its path. Although stream derangement has several of the features attributable to river capture, including barbed tributaries, it is the result of an entirely different machanism. Glaciation is the chief cause of stream derangement. An example of this is to be found in North America where the present course of the upper Missouri River is largely the result of the advancement during the Pleistocene Ice Age of the ice-sheets across the preglacial course of the Missouri. At that time if had its outlet to the north in the Hudson Bay in Canada. The glacial advance caused the upper Missouri to abandon this route and establish a new course, roughly following the glacial boundary, to the Mississippi flowing south to the Gulf of Mexico.

Superimposed Drainage

Some river systems have managed to maintain their original patterns despite obstacles erected during the geologic history of the area. Much of Britain, for instance, was originally covered by chalk and jurassic rocks, lying unconformably above more ancient palaeozoic rocks. As the land emerged from the sea during the late cretaceous period, rivers were formed consequment to this surface.

As the valleys deepened, stripping away much of the younger cover, the rivers cut into the older, more resistant rocks, many with structural features which resulted in barriers across the river courses. In highland Britain, although the chalk cover has disappeared, the rivers remain carved into spectacular valleys through older rock while maintaining a close approximation to their original courses. This type of drainage pattern in which a river system has been let down over an older landscape, is reffered to a superimposed or epigenetic drainage.

If a stream continues through its erosive cycle whilst mountain uplift and folding are in progress it is said to be an antecedent stream, for the river is older than the earth movements. The resulting features of this antecedent drainage pattern will depend on the rate of the river's downcutting. Antecedence has been used to explain the nature of some of the rivers flowing across the Himalayas, notably the Arun and Tista, which flow against the structure of the mountains through impassable gorges.

Working Stages of Winds

It is very interesting that after discussion on River's at work, we now have to describe about the winds at work. Winds is the only physical agent which carry rock particles as against the normal force of gravity, for long distances.

As we know that like river, wind has also an important place to discuss about the work of wind depends on the three agents like:

1. Erosion
2. Transportation
3. Deposition

Erosional Work of Winds

(a) *Deflation*: Loose, dry and incoherent rock particles get uplifted and start blow above the earth's surface due to the help of winds. Such removal of loose particles, only hard and compact rock masses remain in their original places, whcih are exposed for further erosion. It is true to say that a process in which the loose rock particles are removed by the impact of blowing winds is called by the name of deflation. Mainly Rajasthan in India, is best example.

(b) *Abraision*: Lifting up the rock particle by blowing wind and carried by it. In their path of travel, rock particles jump and colliode with each other. Abraision is in broad term as that sometimes the loaded wind with such rock particles attain a considerable erosive power, which helps in eroding the rock surfaces by rubbing and grinding actions and produce many changes, this is called by the abraision. It mainly resulted in two ways:

(i) Nature of the region over which the wind blows.
(ii) Velocity of the wind.

In desertic and semidesertic area, the wind erosion is more effective. It is well-known that desert sand particles consist of a diameter varying between 0.02 to 0.80 mm. As erosive power is directly proportional to the velocity, when the wind is blowing with a high velocity it generally removes dry and loose rock material. Learn winds through given in Figs 15.1 to 15.8.

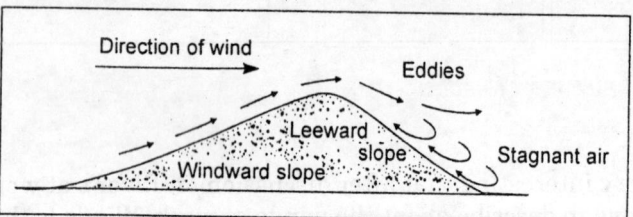

Fig. 15.1: *Direction of wind.*

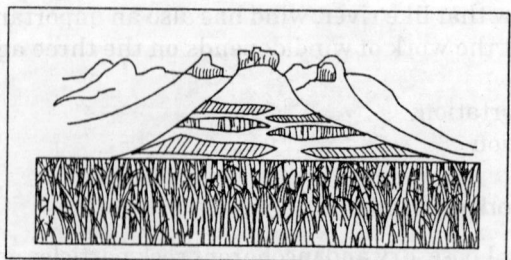

Fig. 15.2: *Inselbergs.*

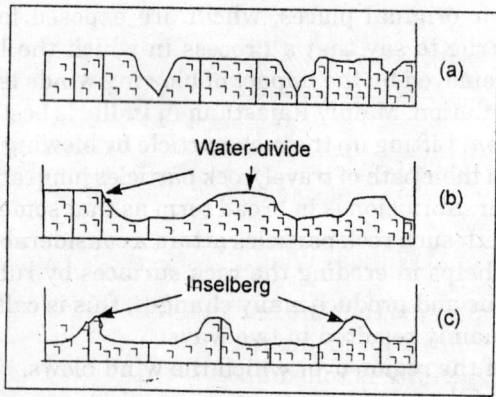

Fig. 15.3: *The development of inselberg.*

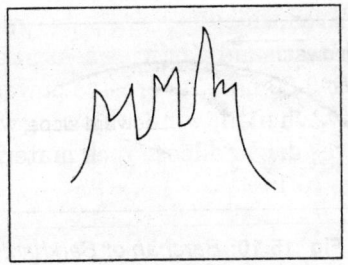

Fig. 15.4: *Needles.*

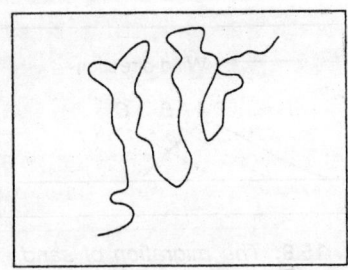

Fig. 15.5: *Castellated chimneys.*

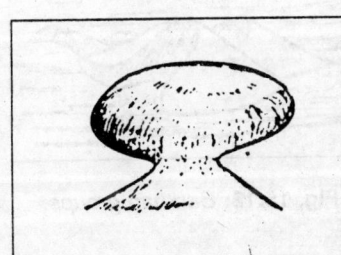

Fig. 15.6: *Mushroom rocks.*

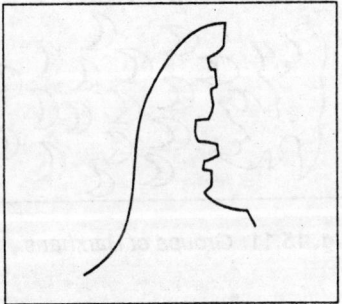

Fig. 15.7: *Demoiselle.*

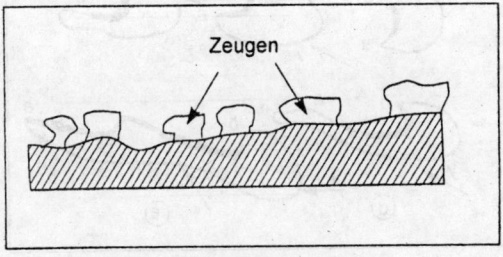

Fig. 15.8: *Zeugens.*

Transportational Work of Wind

It has been observed that the finer and lighter rocks of < 0.06 mm diameter are lifted up in the air and are carried in suspension while the bigger size is rolled upon the surface of the earth. Red rain of February 1903, which fell at certain places in the Great Britain, this was brought by winds from Africa.

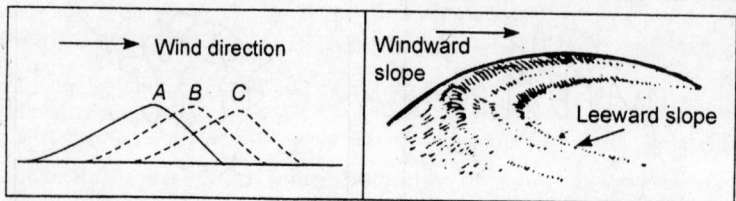

Fig. 15.9: *The migration of sand dunes (From A to C). The successive portions of sand dunes.*

Fig. 15.10: *Barchan or Barkhan.*

Fig. 15.11: *Groups of Barkhans.*

Fig. 15.12: *Barkhan groups.*

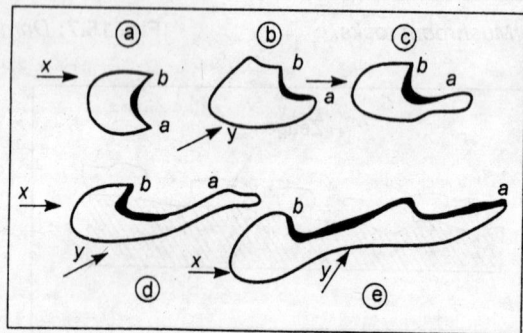

Fig. 15.13: *Development of a seif.*

Depositional Work of Wind

As we know that when the transportational capacity of wind ceases, the depositional work of wind take place, either on the land or into the sea. It is known fact that about 1/6th of the total land area of the world is desert.

Dunes

Sand dunes occur when some obstrution comes in the way of the wind, carried sand particles. Sand particles fall down due to the gravitational pull of the earth. A typical sand dune is a crescent.

The movement of the dune takes place from the windward side towards the leeward side. The action of the blowing wind erodes the sand particles and gradually takes them up to the crest, and lastly these particles are dropped down on the leeward side.

Loess

Wind which carry the fine dust particles and its deposition is called the *loess*. Then we can say that loess is a deposit of fine dust particles away from desert regions, where water and vegetation is in abundance. The term "loess" is originated from China.

16 Works of Lakes

Lakes are a familiar sight in most landscapes, but they are notoriously difficult features to define except in the most general terms. Broadly speaking, lakes are bodies of water enclosed by land, occupying hollows on the earth's surface. They form when the floor of a hollow is relatively impermeable, preventing the water from seeping into the ground, or when the floor lies below the water table the heighest level at which the pores of underlying rock are saturated with water.

Despite this basic common feature, lakes have a huge variety of different characteristics. They may be very large or very small, deep or shallow, natural or man-made, fresh-water or salty, and even permanent or seasonal. They vary in size from the small ponds known by different names in different parts of the world. For example, tarns in England, lochans in Scotland, etangs in France, and billabongs in Australia—to the large features frequently reffered to an inland seas and in some cases given the name "sea". In fact, the Caspian sea in Central Asia is the largest lake in the world, with a total area of 371,800 km^2.

Larger lakes often exhibit characteristics very similar to those of seas or oceans, though usually on a smaller scale. For example, winds blowing over larger lakes can generate waves of sufficient size to result in erosion processes simlar to those at the sea coast. Another similarity between seas and lakes is the creation of deltas where river flow into them and lakes even have a type a tidal movement known as a seiche. Although seiches usually involve rises and falls in lake level of only a few centimetres they occur on a much shorter time-cycle than oceanic tides and may produce over 20 high, "tides" per day.

Although the water in lakes is usually described as fresh, under certain circumstances lake water can have a higher salt concentration than sea water. Thus, in semi-arid parts of the world, rather than being drained by an outflowing river, lakes frequently lose water by large-scale evaporation form their surfaces, leading to the build-up of salts in lake water. The Great Salt Lake in Utah, USA and the Dead sea between Israel and Jordan, e.g. both have salt concentrations six to seven times than that of sea-water.

It is common for a single lake to be due to more than one cause. However, apart from those produced artificially, all these hollows can be sad to be manifestations of the unstable land surface and climate of the earth.

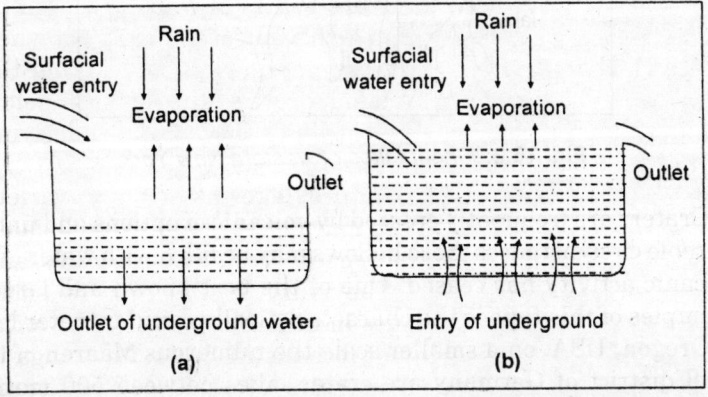

Fig. 16.1: *(a) The outlet of the water of lake in dry climate, (b) the outlet of the water of a lake in humid climate.*

A significant number of lakes are formed as a result of crustal or tectonic movement, volcanic activity, or changes in relative sea-level. Tectonic activity involves subsidence of the earth's crust, a mechanism that was responsible for the creation of Lough Neagh in Northern Ireland and Lake Victoria in East Africa, or down-faulting of a block of land, as occurred in an area of Central Asia to produce the deepest lake in the world, e.g. Lake Baykal. The best known down faulted blocks occur in the rift valleys that extend in a linear belt from Jordon to East Africa.

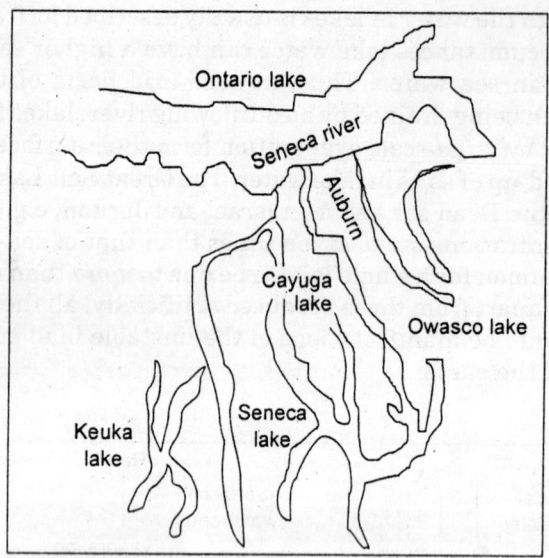

Fig. 16.2: *Finger lakes of western New York.*

Craters are frequently created by volcanic eruptions and under suitable circumstances these hollows may be filled with water after volcanic activity has ceased. One of the best known and largest examples of this type is the characteristically circular crater lake in Oregon, USA, on a smaller scale the numerous Maaren of the Eifel district of Germany are crater lakes between 500 metres and 1 kilometre in diameter. Occasionally, a lava flow or accumulation of other volcanic material may block a valley and so form a lake. One example of this is lake Kivu in East Africa, which originally drained northwards to the Nile but following the eruption of the Birunga volcanic field, has since overflowed southwards to lake Tanganyika.

If relative sea-level falls, either by a rise in land-level or fall in sea-level, or both, numerous lakes will be left in hollows on the former sea floor. This happened on several occasions during the Ice Ages, for at various times sea level fell due to the abstraction of ocean water to form ice, and land-level rose as ice sheets that had depressed the land surface by their weight melted.

A second group of processes responsible for lake formation involves erosion of various kinds. Moving bodies of ice are probably the most important of these erosion agents, leaving a varaiety of attractive lakes in formerly glaciated areas. They form, for example, in deep, rounded hollows, which is called as *cirques* or *corries*, that have a rock lip or in deepened basins in troughs scoured out by glaciers. These are termed ribbon lakes, of which the finger lakes of New York state and the lakes of the Italian Alps Maggiore, Como and Garda, which are good examples.

In low-land areas, glacial erosion has often produced a very irregular topography that becomes a landscape of lakes and rocky out crops. Topography of this type is found in parts of the Canadian sheild, north-west Scotland, as well as in Finland.

Another type of erosion that produces lakes is solution, by which rock material is dissolved in acidic water. In limestone areas this process may form steep sided hollows known as *swallow holes* up to 30 metres deep which can fill up with water if they either lie below the water table or become clogged with inwashed clay. Lough Derg on the river shannon in Ireland is thought to be due to the expansion and deepening of the limestone bed of the river by surface solution.

Deflation, the erosive action of the wind in desert region, can also produce hollow of substantial size. If the hollows are eroded down to a level below that of the water table, they become shallow salt lakes or swamps. Many oasis in the Sahara originated in this way.

There are rare examples of lakes that have developed in the impact craters of large meteorites. The largest positively assigned to this origin is lake Bosumtwi which occupies the Ashanti crater in Ghana and is about 10 km wide.

Barrier Lakes

Lakes involves various types of deposition. Glacial deposit are often laid down in irreglular sheets, in which the numerous hollows may fill-up with water. Two other lake types associated with glacial deposits are those formed behind a barrier of terminal moraine and kettle lakes, formed in hollow produced by the melting of blocks of ice.

River deposits also form lakes. Common examples occur where river deposits seal up both ends of abandoned meander loops, thus forming ox-bow lakes, while the building of natural levees on both sides of building of natural levees on both sides of river channels can prevent flood waters from flowing back into the river channel. Creating flood plain lakes. In coastal locations this action produces delta lagoons, though these are often partly sealed by marine deposits.

Vegetation growth and animal activity can create barriers for lakes, but man is by far the most effective dam constructor. The growing demand for water have led to the large scale impounding of rivers or enlargement of previously existing lakes. In the United States alone over 40,000 square kilometres which is now covered with artificially impounded water, it is generally considered as that an area larger than Belgium.

Transient Existence of Lakes

It is important to mention here that one feature common to all lakes is their transient existence. Not only can their level change from time to time but eventually all lakes tend to be eliminated altogether. Lakes have been observed to drain temporarily after earth tremours, while the tilting of blocks of the earth's crust may lead to a more permanent elimination. Where lakes owe their existence to a position of the water table above the floor of a basin it is clear that a fall in the water table will result in a drop in the level of the lake and in some cases to its complete disappearance.

Just as lakes can be formed by a fall in relative sea-level they can be destroyed by a relative rise of sea-level. This must have happened many times at the end of the Ice Ages, as ice masses melted and returned water to the oceans. The best known example is probably the Baltic sea which at one time towards the end of the last Ice Age was a fresh water body known as Ancylus lake. As sea-level rose its threshold was overtopped and it became an arm of the sea.

Climatic change can produce changes in lake levels. For example, during the Ice Ages, evaporation rates were much lower in many presently semi-arid areas and a mumber of so-called Polluvial lakes existed. The largest of these was lake Bonneville in utah which covered an area of 50,000 square kilometres and was over 300 metres deep in places. The great salt lake is a present day remnant of it. Similarly, most desert lakes or playas are particularly transient features. They may form during an occasional heavy shower but due to rapid evaporation, they quickly disappear.

Lakes may also disappear if the dams impounding them are broken through or removed. Ice dammed lakes, for example, are drained if the water is able to escape under the ice or if the ice melts.

Lakes are of great importance to man. They not only provide natural reservoirs of water, but also help to regulated the flow of rivers, thus preventing excessive flooding and intermittent flow. However, outlet rivers supplied by the overflow from the lake tend to erode down through the barrier holding back the lake waters and may eventually drain the lake.

Another slow process by which lakes eventually disappear is sedimentation or silting up. If none of the previously described processes leads to elimination of a lake, then gradual sedimentation on the lake floor and vegetation growth will eventually fill it in although this may take thousands of years.

One interesting aspect of lakes that have been filled by sediment and vegetation is that they can yield considerable information regarding the geomorphological, vegetational and climatic histories of surrounding areas. Radiocarbon dating of the bottom deposits gives a date for the creation of the lake and in glaciated areas gives a clear indication of when the area was deglaciated. Analysis of the pollen present throughout the core allows scientists to work out the vgetational history of the area and together with the study of beetles and other insects present is valuable in deciphering the climatic changes since the lake was created.

17 Terrains Made by Limestones

As we know that limestones produce one of the most distinctive types of terrain. A variety of features associated with limestone outcrops are recognisable throughout the world; some of these have been known for centuries by local names and as a result words from many different languages make up the current vocabulary describing limestone features. Many such words are either Slovene or Serbo-Croats borrowed from the limestone terrain in Yugoslavia, an area composed of rugged mountains the Dinaria Alps, and the low plateau of Istra. This region is known by the name of Karst, a term which is now commonly used by geologists to mean limestone terrain.

Natural waters, whether in the soil or in a stream, are actively erosive or aggressive when they first come into contact with limestone of karst areas. All limestones, including both marble and chalk, are largely composed of calcite ($CaCO_3$) one of the most common and readily dissolved minerals. As water percolates through the soil it picks up a considerable amount of carlondioxide from the high percentage found in soil air. Calcium carbonate reacts with carbon dioxide and water to form the very soluble salt, calcium bicarbonate, which may be carried away in solution.

Limestones are mainly dissoloved in two ways. Firstly, water percolating through the soil cover gradually lowers the over all land surface. In humid, mid latitude areas, the rate at which the ground surface is lowered may be as much as 50 to 100 mm in a thousand years. Secondly, water which has converged into surface streams over non-limestone rocks concentrates solutional activity at the point where the river passes on to a limestone outcrop.

Another important feature of limestones is the fact that they are well jointed. Joints are fractures in which little or no movement parallel to the walls of the fractures has taken place. Vertical joints originated due to contraction following the drying out of the limestone when it was first deposited. It the joints are horizontal, they mark changes in the character of the sediment or pauses during its deposition, and are termed bedding joints or bedding planes.

Finding a path through the limestone by water is due to joints. Some features of karst areas are due to essentially to jointing and are very similar to comparable landforms in other well jointed rocks which are non-soluble. For example, gorges cut into the rock are often flanked by sharp sides and steep cliffs that follow the outline of the jointing pattern. If a limestone is less well jointed, like the porous chalk outcrops, smoother outlines develop in the landscape chalk is a soft, white type of fine grained limestone. Because of its purity-chalk is composed almost entirely of calcium carbonate it is highly soluble and easily eroded, giving rise to the gently rolling scenery found in the hills of eastern and southern England and the hills of Picardy and Artis in France.

It is the combination of these properties of limestone, being both soluble and well jointed, that explains the distinctiveness of karst area. For example, a surface stream may disappear at the junction of major joints creating the limestone feature known as a sink or swallow hole. In England, a well-known example of a swallow hole is gaping Ghyll, a deep opening in the carboniferous limestone in the Yorkshire Pennines. If a disappearing stream continues to erode its bed up stream from a swallow hole, the valley ends abruptly in a horseshoe shaped cliff above the opening. This is term a blind valley. The valley of the River Reka in Yugoslavia ends in this way.

Underground Cave System

Once underground, the stream continues to eat along the joints, enlarging caverns in the limestone. In horizontally bedded strata, caves are made up of a series of vertical drops or pitches linked by horizontal passages. In mountain karst, the total depth of caves is considerable. For example, near Lyon in eastern France, Gouffre Jean Bernard is estimated to be 4000 ft deep.

Just below the earth's surface, beneath a soil cover, solution slowly etches out the joint pattern. Due to deforestation and the consequent soil erosion in many areas in the Middle Ages, roughly rectangular blocks known as limestone pavements which is now revealed as a feature of well jointed limestones. In north-west Yorkshire, the pavement is called a clint and the soulion enlarged joint is termed as a grykes. The German word Karren is widely used to describe grykes and solutional gutters on bare limestone outcrops.

Where major joints intersect solutional activity may tend to concentrate and develop a funnel shaped surface cavity, now widely described by the sloven word doline. Dolines may be as much as 330 ft in diameter. If the base intersects an underground drainage system, the dolines sides may collapse to form a circle of sheer cliffs, termed a shaft doline. Such a doline, broadened to afford a view of the underground river emerging from one cave them disappearing into another, is called *karst window*.

However, shaft dolines and karst windows are not common. Cave systems, often have no noticeable effect on landforms developed above them on the surface, although some authorities suggest that a number of valleys in karst areas have formed by caverns collapsing.

As we know that drainage patterns, however, are markedly affected by cave systems, in many cases entire valleys are now streamless because all precipitation finds its way into underground fissures, voids or caves before it can concentrate as a surface stream. Such dry valleys can therefore be attributed to a lowering of the underground water levels, as solution progressively enlarged joints which were once very narrow.

Dry valleys are also partly explained by climatic changes during the quaternary and post-glacial periods of the last two million years. For instance during the quaternary period, the Karst in Yugoslavia was under the strong meterological influences of the subpolar front and the ice cap of the Alps. Cyclones from the sea brought great quantities of moisture to the karst mountains, much of it falling as snow, and the enormous floods of summer melts exceeded the capacity of the cave systems to lead all the water underground. This abundant surface flow over the permeable limestone cut out valleys, which today are dry. Dry valleys and a lack of surface drainage are characteristic of most chalk country.

One explanation suggested for the formation of streamless valleys in the highly porous english chalk area is that they were cut when the ground was permanently frozen. Which meant that the limestone was temporarily impermeable.

Climate is an important factor when present day comparisions are made between the main karst areas of the world. Because solutional processes are governed by the amount of effective precipitation that falls, and by temperature, limestone terrains are not the same in different climatic zones. For example, in arctic or alpine areas, the main agent is the large volume of snowmelt which runs over bare rock surfaces in late spring and summer. Karren are therefore distinctive characteristics of such regions.

In contrast, warmer karst area are dominated by large depressions known as *poljes*, from the Serbo-Croat word for field. A polje, however, is a very unusual type of field. It is the very flat floor of a large, isolated karst depression, cut accross the bed rock and bounded by steep slopes on all sides. Poljes, which are often several kilometres long, experience seasonal flooding. Winter run-off from the surrounding hills is too great to be drained off by the existing cave systems. So a large temporary lake forms. Swallow holes at the edge of poljes are called *ponors*. Some underground system act as springs in winter when water is plentiful and swallow holes in summer. These seasonally fluctuating features are termed estavellen.

Poljes have not developed in cooler environments, although the turloughs of country clave in western Ireland are very similar. The slopes of the perimeter of high land around the turloughs are gently inclined, whereas in humid tropical regions the contrast between polje floor and the surrounding hills is more accentuated. The major reason for this is the increased acidity of water in warm, moist areas, caused by the swift decay of dead vegetation on the ground. Highly acidic water subjects the limestones to rapid erosion.

In karst country in the humid tropics where the hills have become isolated in a flat plain, the distinctive tower karst country is developed. The junction of the polje floor with the karst tower is very abrupt and there is often solutional undercutting of the tower. Rock falls at such points have been recorded in recent years in Malaysia. Another distinctive feature of tropical karst is the enlargement of adjacent dolines until any flat land between them is eliminated. The higher land is reduced to a series of isolated peaks or cone karst. The classic example of cone karst is the cockpit country of Jamaica.

Much of the calcium carbonate dissolved from limestones remains in solution untill the drainage waters reach the ocean. But in certain circumstances, deposition of calcium carbonate may occur in sufficient quantities to create new landforms. Deposition seems to happen more in warmer climates and at points where water is slow-moving or running as a thin film over rock. Precipitated calcium carbonate is called travertine and occasionally forms water-fall screens, which look like petrified water-falls. Within the tropical belt the best known limestone terrain, the coral atolls, are formed by the abstraction of calcium carbonate from sea water by coral polyps.

There is always an impermeable basement to any limestone block which will halt the downward percolation of ground water. Within the limestone strata, thin impermeable layers like shale, volcanic ash or marl acts as the outlet of a spring or resurgent stream until the layers are eventually breached. In a covered Karst, a non-soulable or impermeable stratum remains above an underlying limestone mass. If the cover is relatively thin and the limestone strata beneath undergoes extensive solution, they may collapse, creating dolines in the cover rock itself.

18 Working Stages of Glaciers

As we know that about ten percent of the earth's land surface is covered with ice. As recently as 70,000 to 10,000 years ago, during the last major advance of Pleistocene Ice Age, almost 30 percent of the earth was blanketed by ice and snow. Profoundly affecting the landscape of many areas. Today, the erosional effects of ice are reshaping such spectacular scenery as the mountains and valleys of the European Alps.

The ice which feeds the world's vast ice sheets and glaciers primarily forms through prolonged accumulation of snow. However, fresh snow is light and easily disturbed and before it is converted into heavier, more tightly compacted ice, various processes have to take place. The change is mostly due to the compression of snow by the weight of successive additions from direct snowfalls and from avalanches down surrounding slopes.

Infact, in some areas summer temperatures and the increasing pressure of successive layers of snow fall cause some of the ice particles to melt. The melt water produced then trickles down through the snow and refreezes so that the compressed snow is

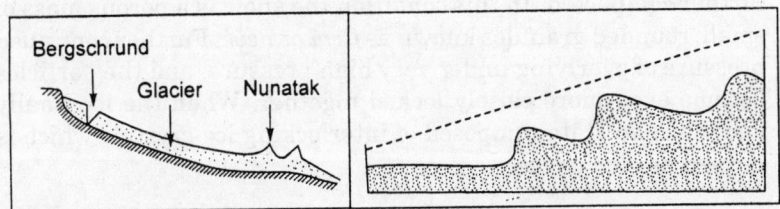

Fig. 18.1: *Profile of a glacier.*   **Fig. 18.2:** *Long profile of a glacier.*

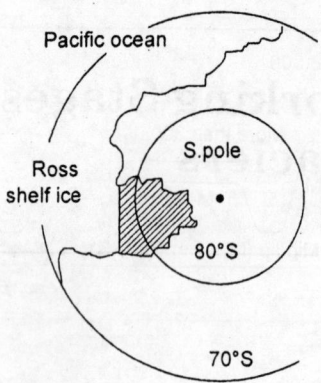

Fig. 18.3: *Rose-shelf ice.*

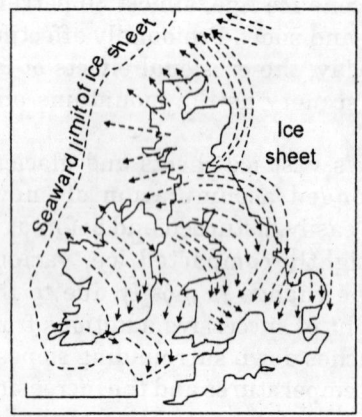

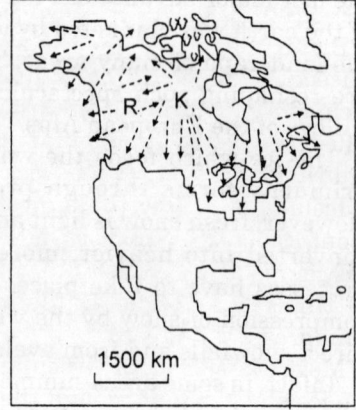

Fig. 18.4: *Advance of ice sheet in the British Isles.*

Fig. 18.5: *Advance of ice sheet in America.*

further compacted. In this condition the snow is a porous mass of small, rounded granules known as *firm* or *neve*. Further continued pressure of overlying under very high pressure, and the particles become even more closely locked together. When the ice finally forms a solid state composed of interlocking ice-crystals which is termed as *glaciar ice.*

TYPES OF GLACIERS

There are mainly two types of glaciers:

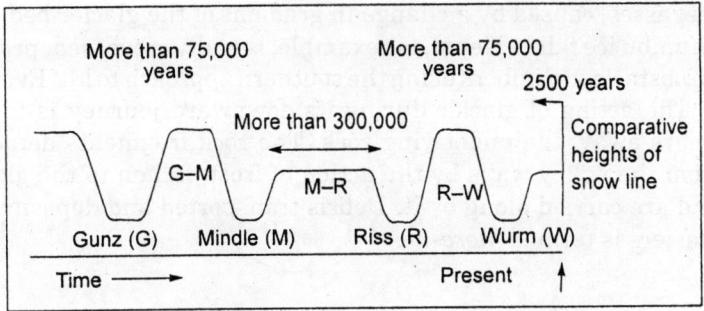

Fig. 18.6: *The extent of ice ages.*

1. Valley glacier
2. Continental glacier

By far the most extensive are continental glaciers, vast areas of ice spreading out under their own wight. The larger spreads, such as those covering Antarctica and Greenland, are generally called *ice sheets*, while similar areas of ice, such as those in Iceland and Norway, are known as *ice caps*.

Iceland is having the largest European ice cap, which covers about 19,500 km² (7,500 sq. miles) and is about 230 m thick. When it is compared with the Antarctica ice sheet, which covers an area 10 times the size of England and Wales, this is relatively unimpressive.

Most of the area of Greenland is permanently covered in ice, the ice sheet spills through the coastal mountains and descends in the form of outlet glaciers.

Valley glaciers are slow-moving rivers of ice which flow downwards from higher to lower land as valley. They can be devided into three sections, starting with the source or "zone of accumulation" where accumulated snow is transformed into glacier ice.

The movement of mass of ice out of the basin enters its zone of transit. As it flows downwards, its surface begins to spit and crack, probably due to movement within the ice mass itself, to form openings or crevasses.

The margins of the glacier is the occuring place of crevasses. It is due to the difference in the rate of movement of ice in the middle of the glacier and along the valley wall. Other features of this stage are steep ridges called *ceracs*, and ice falls, masses of ridges and crevasses, caused by a change in gradient of the glacier bed. The khumbu ice fall in Nepal, for example, is badly crevassed, proving an obstacle to climbers using the southern approach to Mt. Everest.

The acting of glacier during its downward journey is that it wears away, the underlying rock. The rock fragments detached from the valley walls by the action of frost, fall on to the glacier and are carried along by it. Debris transported and deposited by glaciers is termed *moraine*.

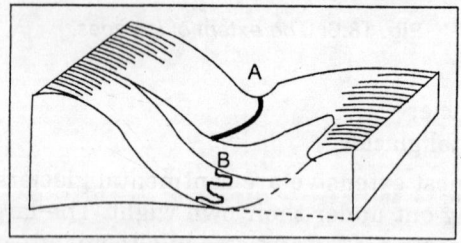

Fig. 18.7: *The formation of U-valley; A = Initial, B = U-valley.*

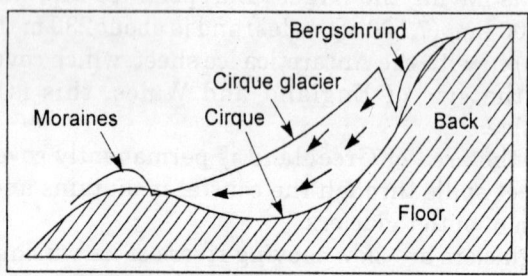

Fig. 18.8: *Morains, cirque, etc.*

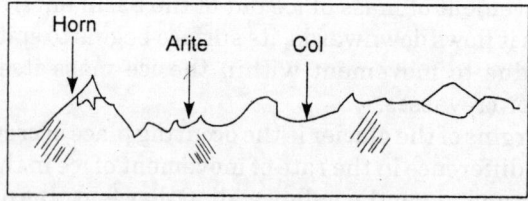

Fig. 18.9: *Horn, arite, col, etc.*

At the final stage, the ice mass moves beyond the snow line (the lower limit of permanent snow) into an area of decline or zone of ablation. At this place, the glacier is subject to the effects of evaporation and melting which contributes to this decline.

EROSIONAL WORK OF GLACIER

Glacial erosion forms the first part of a continuous chain of glacial activity, which ends with transportation of debris to its place of deposition. There are various factors which contribute to the degree and rate of erosion; the strength, hardness and pattern of joints and fractures in the underlying rocks, the thickness and speed of the glacier, and the concentration and characteristics of the rock fragments within the base or sole of the ice.

Three key processes are involved in glacial erosion.

1. Abrasion
2. Crushing
3. Plucking

Abrasion refers to the scraping or scratching action on the glacier bed of the rock fragments held in the lowest levels of ice.

Crushing involves the grinding or breaking down of bed rock by the weight and pressure of ice and debris.

Plucking is the direct incorporation of crushed or abraded material into the glacier base.

The process of crushing and plucking are sometimes collectively known as *quarrying*.

The fourth unimportant process involves erosion by meltwater beneath the ice, concentrated into channels or cavities. Such fluvial erosion may constitute and important method of entirely removing loose material from a glacier basin.

Bed level abrasion is principally controlled by the pressure of overlying ice. As the ice-thickness increases, so the pressure at the base of the glacier is also increases, resulting in accelerating rates of abrasion. It appears, however, that beyond a certain ice-load rates of abrasion begin to decline and may be eliminated altogether.

Crushing of bedrock to produce rock fregments and boulders is also dependent upon the weight of the load of ice. Here too, however, it is thought that as ice thickness increases so the degree of crushing will reach a peak and then decline.

The action of plucking requires that ice comes into direct contact with loosened materials. These material are then incorporated into the mass of ice in two ways fragments may be picked up and encased in the viscous flow of the ice, or a mixture of water and rock debris may amalgamate by being frozen on to the glacier sole.

Particular type and extent of these processes depends on temperature at the glacier bed. The temperature at the sole of wet based or temperate glaciers and ice caps are at the pressure melting point due to the sheer weight of overlying ice and warmer mean annual temperatures at their surface. A thin layer of water may consequently exist on the bed, causing the ice mass to move faster. For example, most valley glaciers in the Alps, in Scandinavia, in parts of the rocky mountains and in ice land are wet based. The whole range of erosional processes abrasion, quarrying and melt water beneath the ice are vigorous in these glaciers.

Cold ice mases or dry based, on the other hand, characterised by temperatures at their base below the pressure melting point. In other words, they are frozen to the rock and no water can exist at the bed. Such conditions occur over large areas of the Antarctic and Greenland ice sheets and in some valley glaciers in high latitudes.

Naturally, the action of melt water is completely suppressed here, and although plucking of materials appears to be somewhat enhanced under cold conditions, in general erosion if less active.

LANDFORMS PRODUCED BY GLACIAL EROSION

The glacial erosion is creater of a great variety of distinctive landforms. On a small scale, scratches known as striation are produced mainly by abrasion. Rock fragments, held tightly in the glacier base, are dragged across the bedrock cutting linear scratches of varying depth, i.e. from a few millimetres to a metre or more and up to 100 metre long. Gouges, chatter marks and crescent shaped cracks, close series of gouges, also originate from boulders coming into contact with the bed and causing it to fracture.

On a large scale, ice may mould hummocks, mounds and small hills of hard rock into asymmetric, streamlined landforms. Such features, termed roche mountainees or whalebacks, which are found widely in glaciated regions such as Scotland, North Wales and Scandinavia.

In mountanous regions, such as the Alps and Rockies, hollows or cirques are found above and at the heads of many glaciated valleys. These deep bowl-shaped rock depressions with a steep backwall and a flatter floor frequently contain a small lake or tarn once the ice has melted. Many cirques have diametres up to 1–2 km, with cliffs up to 1,000 m in height.

Glacial valleys are distinguished by their generally straight, deeply hollowed appearence and constitute one of the most spectacular elements of glacially seulptured highlands. In cross-section their shape is broadly that of a "U" contrasting strongly with the "V" shape of valleys produced by fluvial erosion. Never the less most glacial troughs have evolved due to the modification of former river valleys occupied by temperate glaciers during glacial episodes.

Tributory glacial valleys are frequently left as hanging valleys above a main glacial trough since glaciers, unlike rivers, do not combine to erode to a common base level. Each individual glacier acts as a semi-independent force, and the main, usually thicker glacier erodes its valley more deeply than the tributaries.

Fjords are valleys reaching to sea level in mountainous coastal localities, created either by glacial erosion below sea level, or by the submergence of glacial troughs as the sea level rises. They usually have excessively deepened basins and precipitous sides, the result of vigorous erosion due to considerable ice thickness and high ice velocity. The skelton inlet in Antarctica has a maximum water depth of 1,933 metres and a full vertical elevation range of 4,500 metres. Other equally impressive fjords occur in Western Norway, Greenland and Scotland, and on the north west coast of North America, in chile and in South Island, New Zealand.

Some elemenls of mountainous glacial landscape are the combined result of a variety of erosional processes. Aretes for instance, are the knife edged, serrated ridges that separate glacial valleys or cirques. Pointed pyramidal peaks or horns, such as the matterhorn and weisshorn in the Alps, form as a result of erosional retreat of a series of cirques around a central mountain massif.

In contrast to highland regions, the topography produced by glacial erosion in lowland areas is less marked and exhibits quite different characteristics, which are related to ice sheet activity rather than that of valley glaciers. Frequent among landforms here are ice-scored plains made up of numbers of rounded rock-knobs and hollows, often termed knock and lochan topography in Southern

Scotland. In some locations ice has imprinted a dominant grain or lineation on the scoured terrain, which mirrors the overall direction of ice flow and is called *fluting*. Large tracts of the Scandinavian and the Canadian shields display this kind of landscape. On a greater scale, extensive basins may be formed by glacial erosion, especially near the margins of former ice sheets. A key element governing the glacial processes which produce these types of landform is the condition of the bed rock, and the relief of undulating lowland regions, exposed when the ice melts, may closely reflect varying degrees of abrasion and quarrying of weak and resistant strata.

DEPOSITIONAL WORK OF GLACIER

Glaciers are also very active in forming various landforms during its depositional work. Infact, during the last two million years, over 30 million square kilometres of North Europe, North America and Asia were periodically covered by great ice sheets. The melting of the ice left large expanses of these continents covered by a veneer of glacial sediments, varying from a few centimetres to more than 400 metres thick.

On land, glacial sediments are now of great economic importance. They constitute some of the world's most fertile agricultural land and are a major source of sand and graval for building aggregate, i.e. the material added to cement to make concrete. On the deep sea floor, glacial debris produced by the melting of ice bergs, covering vast areas, has lain relatively undisturbed since deposition and thus provides crucial clues to the recent glacial history of the earth.

The great deposition of glacier forms the final part of a train of glacial processes beginning with the incorporation of soil and rock into a glacier as it eroded the land. These materials are then transported and often modified by moving ice and water before their final deposition in a range of environments, often producing distinctive landforms.

The sediments of glacier are classsified into two main groups:

1. Till.
2. Sorted deposits.

Till is a random mixture of rock particles, including at times both tiny clay particles and massive boulders; it is deposited directly by the glacier.

Sorted deposits require the action of water or wind to sort their particles according to size and weight. Once deposited, sediments also altered by chemical processes, which adds to the distinction between younger and older material.

Deposits of till produced by the ice are generally described as moraine, a term which is also applied to all debris transported by a glacier.

Till can be transported by ice in various ways. It is carried at or near a glaciers surface, elongated lines of sediment may form, either at the sides of a glacier in the ablation or melting zone later at moraine or where laterals form tributary glaciers coalesce in the centre of a main glacier (medial moraine). If sediments are carried at the bed of a glacier, they may form closely stacked layers with occasional lumps of other materials. The thickness of these basal layers is about one percent of the glacier's depth and the concentration of debris decreases upwards from the base. As it is carried along, till is often modified as pebbles are broken, scratched, smoothed and frequently changed in shape. Only a very small percentage of material survives such destructive processes for more than about 30 km and then only in a ground-up form. Granite and metamorphic rocks, for instance, may be reduced to fine sand over long distances, while shales persist only as clay sized particles. However, very large boulders may be transported many kilometres by ice and, when deposited, stand out as erratics, rocks or rock fragments foreign to their new geological setting.

Till carried at the base of a glacier is often deposited in horizontal sheets. Sediments close to the glacier sole are continuously transported by the force of the moving ice until this force becomes insufficient to carry the debris with it. The immobile till is then described as lodged. Basal till is also deposited by melting out, which occurs when the lower layers of a glacier melt, leaving their sediment on the bed.

Till can also originate from sediments carried on the surface of a glacier. As the surface melts, a thick deposit is produced on top of the glacier, heavily charged with meltwater; this is known as *flowtill*.

Landforms Produced by Till

Landforms produced by till, after the ice masses have disappeared, is that of a shapeless sheet of boulder clay, such as those found in

Eastern England, Northern Germany or parts of the Eastern USA. Such sheets are usually the product of lodgement and sometimes carry a superficial layer of flow or meltout till. Within spreads of the till, however, a variety of other forms occur, related to special conditions on the glacier bed during deposition.

Drumlins are streamlined mounds of till, commonly appearing in clusters. They may have rock cores and often occur behind moraines formed at the front of an ice sheet. The drumlins which extend over much of Northern Ireland, are one of the largest groups in the world. They vary considerably in size, ranging from a few metres in length and height, to over 1.5 km long and 60 m high. It seems probably that drumlins originated as masses of debris lodged beneath the glacier and subsequently streamlined by moving ice other moulted till forms such as flutes, i.e. ridges parellet to the former direction of ice flow, are produced by similar processes.

Both basal and surface till are often continuously deposited at the front of a glacier or ice sheet. While the ice front remains fairly stationary, materials falling down the front or carried along by meltwater build up a ridge of debris or terminal moraine along the margin of an ice-sheet or at the end or smout of a glacier. Only terminal morain from the last or farthest glacial advance usually remain, but those left after the retreat of powerfully erosive glaciers can be most spectacular. The largest terminal moraine in Britain is the Cromer Ridge near the coast of Norfolk, which is 8 km long and over 90 m high.

SORTED SEDIMENTS

Sorted sediments, often with well-developed layering or bedding, have very different characteristics from till. They accumulate either close to an ice-mass (they are termed ice-contact deposits) or at a greater distance beyond the ice front.

Some of the most important groups of sorted sediments are those deposited by glacial meltwater. Where these are of the ice-contact variety, a distinctive morphology and sediment texture may be produced. Eskers are long, narrow ridges which snake sinously up and down over gently undulating terrain, often reaching sereral hundred kilometres in length and hundreds of metres in height. In some cases they are made up of mounds joined by low ridges, in others of complex networks of criss-crosssing ridges.

The sediments forming eskers are sorted. Pebbles and cobbles, often with little sand and silt, crude stratification is present with cross-bedding and structures such as ripples and dunes.

Eskers are primarily formed by streams tunnelling beneath, within or on a glacier, carrying sediment which sinks to ground level when the icemelts. Observations have indicated that some eskers may also be produced from sediments carried by meltwater streams flowing into pro-glacial lakes. Some of the best developed eskers are to be found in the lake country of Finland and other examples occur in central Ireland and parts of Northern Canada.

Other ice-contact deposits include kame terraces which are laid down mainly by streams flowing in a trough between valley glaciers and their valley walls. As the ice-retreats, deposits are left on hill sides, which sometimes slump down into the valley.

DEPOSITION BY MELTWATER

Pro-glacial sediments, washed out of an ice mass by meltwater, are deposited beyond the ice front and form fan-shaped spreads of sediment called *outwash* where confined in a narrow valley, outwash deposits are called *valley trains*. The condition of meltwater and debris issuing from a glacier determines the nature of the sediments in these trains. Close to the glacier, they may be coarse but decrease in size down the valley.

Where sedimentation is not confined by a valley, outwash fans may coalesce to produce broad alluvial expanses or outwash plains. Much of the south coast of Iceland is made up of outwash material, called *sandar*. In New Zealand, the center bury plains on the east-central side of south Island constitute an outwash plain of over 10,000 km^2 left by sediment-laden meltwaters from glaciers in the southern Alps.

Where glacial meltwaters are blocked by rock-bars, morains or other obstruction, pro-glacial lakes may form and sediments in these are termed glaciolacustrine, i.e. deposited in glacial lakes. As streams enter the lakes, their speed is checked and part of their load is deposited, creating deltas.

CHRONOLOGY OF GLACIERS

Within lakes, finer materials also settle, forming varves, distinctive banded deposits. Each varve consists of two layers, one coarse and light in colour, the other darker and finer. During the summer thaw, the flow of water is able to carry and deposit the coarser material, while in winter, when water movement diminishes due to freezing, the finer particles settle, thus each varve represents one year's deposit. Exceptionally, well-preserved varves found in Southern Sweden were investigated late in the nineteenth century be Baron de Geer. His pioneering and painstaking work of counting varve "couplets" gave some of the earliest accurate dates to the fluctuations and overall retreat ice in Europe.

During the Pleistocene Ice Age, strong winds were generated by the climatic conditions. Fine materials were picked up and carried over many hundreds of kilometres, eventually to be deposited as loess. This is a porous, crumbly silt deposit, usually without bedding, coloured yellow, orange or brown due to staining by iron minerals. Loess occurs in sheets many metres thick, particularly in Central North America, Eastern and Central Europe and China, and it is highly fertile. The loess zone of Europe is one of its most important agricultural areas.

DEPOSITION OF GLACIAL SEDIMENTS AT SEA

The most extensive area of glacial deposition is to be found, not on land, but on the sea-floor. Glaciers descending from coastal mountains may reach the sea and float out if the water is sufficiently deep. The edges of continental ice-sheet may also terminate as a thick shelf of ice over the sea. In both cases chunks of ice may regularly break off at the edges to form ice bergs. These are then carried by currents and wind into the open sea where they eventually melt-away.

Any sediments carried by these icebergs will then be deposited as a mantle on the sea-floor. These glacial sediments contrast strongly with normal marine deposits because of the greater size of particles, their lack of fossils and the often distinctive scratches, cracks, grooves and polish on indvidual grains. Extensive accumulations of such deposits in the southern ocean today completely girdle the ice-covered. Antarctic continent and date back 25 million years. In the Northern Hemisphere they cover the floor of many sea-areas, testifying to the presence, in the not-too-distant past, of ice sheets on adjacent land.

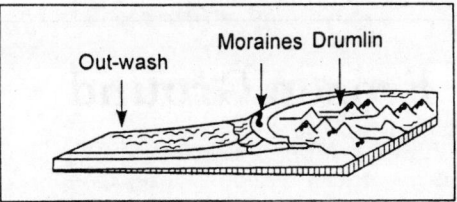

Fig. 18.10: *Outwash, moraines and drumlines.*

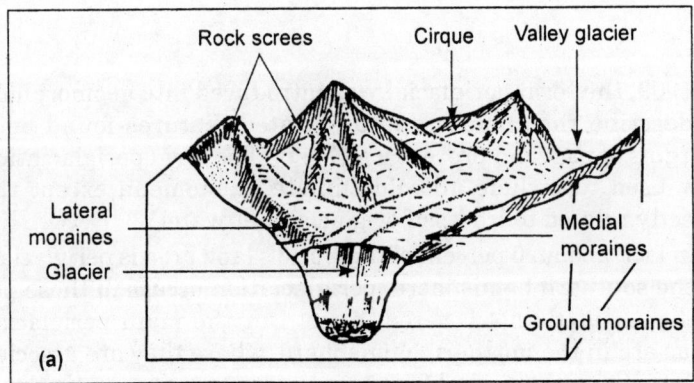

Fig. 18.11: *Types of moraines, cirque, rock flour, etc.*

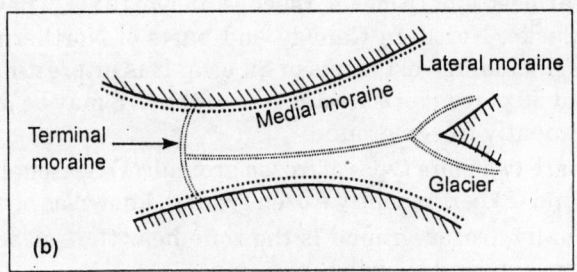

Fig. 18.12: *Type of moraines.*

19 Frozen Ground

In 1909, the term periglacial was introduced into geomorphology to describe the climate and associated features-found on the margins of past and present ice-sheets. The term periglaciation is now used to include all cold climate phenomena except those directly related to ice itself or glacial meltwater.

In fact, about 20 percent of the world's land area is periglaciated. In the southern hemisphere, periglaciation occurs in those parts of the Antarctic not covered in ice. But the main periglaciated areas are in the northern hemisphere, where they are associated with the freezing peaks of many mountain ranges, Arctic Canada and Greenland, and the cold Tundra of high latitudes. Tundra areas are barren treeless plains where the average temperature of the warmest month never reaches above 10° C. They stretch across Alaske, Northern Canada and parts of Northern Europe and Asia, including vast tracts of Siberia. It is interesting to note that about 50 percent of the total area of the CIS may be underlain by permanently frozen ground.

There are two main types of frozen ground: (1) Seasonally frozen ground, and (2) permanently frozen ground, known as *permafrost*.

Seasonally, frozen ground is the zone near the surface where annual freezing and thawing takes place.

In summer when the seasonally frozen layer melts it becomes very fluid. The water in this layer known as the *active layer* if it is underlain by permafrost is unable to drain through the frozen ground and creates a very muddy surface layer which may gradually begin to flow down-hill. This slow, viscous flow is known as *solifluction* and often takes place in the form of huge soil toungese called *solifluction lobes* which are distinctive features of many hillsides in Tundra areas.

In winter, when this active layer refreezes, two other related processes may come into play. These are known as *frost heave* and *frost thrust*. The surface of the ground freezes first and the fraust gradually penetrates downwards towards the deeper permafrost. For a time the intervening layer remains unfrozen. But it becomes subjected to considerable pressure as when of water freezes it increases in volume by about nine percent. Thus, as the ground gradually freezes, it expands by thrusting horizontally and heaving vertically between the solid permafrost and rigid surface layer. The tremendous pressure is most easily released by upward heaving on the surface, producing contortions known as *involutions* in the soil layers. The rigid surface becomes disrupted and objects such as stones, trees, fence posts and even telegraph poles have been seen to be completely heaved out of the ground.

When this spikes of ice grow in the active layer as it freezes in winter, surface displacement occurs on a smaller scale. These spikes of needle ice consist of ice crystals which usually range from one to three centimetres long, but can significantly alter the landscape; they can lift soil particles and stones above the surface and, if this occurs on a slope, when the crystal melt the material will fall and roll downhill.

A further example of the powerful action of freezing and thawing is *frost shattering*, one of the most intense forms of rock fragmentation on earth. When water freezes in cracks, joints or other lines of weakness in rock, its nine percent increase in volume exerts considerable pressures. The expanding ice acts like a wedge driven into the rock, and this ice or frost wedging results in splitting and shattering of the rock.

Frost shattering varies greatly with rock type. Sedimentary and other relatively soft, porous rocks have a great water-holding capacity and are usually well-bedded. Consequently, such rocks are generally more susceptible to frozen wedging than igneous or metamorphic rocks. However, the cleavage planes in metamorphic rocks such as slate and schist provide lins of weakness where frost shattering may occur, producing flat, angular slabs of rock.

In permafrost areas, a variety of features resulting from the melting of ground ice resemble those found in limestone karst areas. These include such features as caverns, disappearing streams and troughs, known collectively as *thermokarst*. Thermokarst is created when the permafrost is disrupted by large-scale climatic changes or by local environmental disturbance.

In the early 1920s, forest was cleared from land near Fairbanks, Alaska (USA), to prepare the area for agricultural purposes. The area had been underlain by vertical wedges of ice and, when the vegetation was removed, the ice-wedges began to thaw, causing the soil to collapse into them. An extensive pattern of depressions and thermokarst mounds developed, varying from 3 to 15 metres in diameter and up to 2.5 metres high, creating an undulating topography quite unsuitable for its intended agricultural use.

The most spectacular features of periglaciated areas are pingos, an Eskimo word describing scattered, isolated hills. The term was introduced into geomorphology in 1938 to describe the conspicuous domes with radially cracked summits that are particularly common in the continuous permafrost zone. They can be over 50 metres high and up to 600 metres in diameter.

As we know that there are two types of pingo; closed and open system types. The *closed system* has been best examined in the Mackenzie River delta area of Canada where 98 percent of the 1,380 pingos mapped occur in fairly level, poorly drained sites such as former lake basins.

The *open type* of pingo is best developed in East Greenland and most common on slopes rather than on level areas. It appears to be associated with artesian springs which provide a continual supply of ground water under pressure. As this water approaches the surface it freezes, and the expanding ice exerts considerable upward pressure on the ground above.

It is interesting to mention that one of the most striking and controversial features of a periglacial landscape is the unusual geometric patterning found in the soil. The most common shapes are circles, polygons and stripes so perfectly arranged as to appear deliberately laid out; yet these patterns occur naturally in the cold terrain. The circles are stony areas measuring from a half to three metres across, often with vegetation growing around their margins. In some the stone are smaller in the centre of the circle than towards the edge.

The polygons are usually larger, sometimes reaching over 10 metres in diametre. They are delineated by vegetated furrow or lines of stones enclosing areas of finer debris. Polygons are best developed on fairly level areas. Where the ground steepens, stripes from these are thought to be the downslope extension of polygons, an intermediate form is the extended polygon or net.

Patterned ground has been the subject of considerable debate in geomorphology, its origin remains unclear, although some polygons seem to result from ice-wedging.

The Habits of Wind, Water and Snow

Violent winds sweep over the cold, exposed periglacial terrain, leaving a trail of wind-scoured features and transported sediments. In fact, wind action is one of the most important processes at work, capable of sand-blasting the rocks to produce sharp-edged, polished stones, termed ventifacts or dreikanter. The thick deposits known as loess are fine-grained sediments swept up and carried sometimes great distances by the wind from bare areas of glacial and outwash deposits.

River flow in the periglacial zone is characteristically irregular. Water will cease to flow altogether during prolonged periods of sub-zero temperatures, but during the spring thaw powerful torrents rush over the river beds causing considerable erosion and movement of debris. Similarly, coastlies in peri-glaciated areas are usually frozen up during winter, but in summer intense shattering of coastal rocks may take place as each tidal rise fills cracks with sea-water which freezes when exposed to air at low tide.

Erosion also occurs in areas apparently protected by a blanket of snow. Beneath patches of snow continual thawing by day and freezing by might leads to bed-rock weathering, with the loosened particles being washed out by meltwater. Snow-patch erosion or nivation thus tends to create large, distinctive hill side hollows that may eventually become cirques.

Homo sapiens, i.e. Man in Periglacial Areas

Homo sapiens, i.e. man faces many problems as he penetrates into periglacial areas to exploit their mineral reserves. The difficulties stem largely from the presence of permafrost and its susceptibility of melting when disturbed by human activities. When the ground is frozen it has great bearing strength but when thawed it turns into mud with no strength at all. Consequently, it is impossible to build centrally-heated dwellings directly onto the permafrost. A common solution is to raise the building off the ground by using stilts or piles so that air can circulate between the heated building and the permafrost. However, the piles are open to attack by frost heaving and must be lubricated and placed sufficiently deep to prevent this happening. Where roads and airport runways are constructed, the ground is insulated by spreading a gravel blanket about one metre thick over the ground.

The provision of services, particularly water and sewers, to polar settlements creates many problems.

In summer there are a few difficulties, but in winter water is in short supply and chunks of ice must be melted down. A number of the larger settlements have piped-in water supplies but the problem is in preventing freezing.

Sewage is often hauled by bucket or wagon and dumped on the river ice to be carried away with the spring break-up. Some settlements have heated sewage disposal systems but it is not possible to put the pipes underground since the heat would disrupt the permafrost. This is also true of the pipelines constructed in such cold, remote areas as Alaska to carry the relatively warm oil from polar oil fields to more temperate latitudes. Buried pipes can be fairly effectively insulated. As alternative is to construct the pipelines above ground on piles, but this is expensive and may inhibit the migration of such animals as caribou to their traditional breeding grounds.

20 The Pleistocene Ice Age

There have been several ice ages in the earth's history, periods when for one reason or another ice caps and glaciers became much more extensive than usual and large areas of the earth's surface where dominated by glacial and periglacial processes. The evidence of these ice ages is there to be seen in the rocks and, although the further one goes back in time the fainter the evidence becomes, we can be certain that there was an ice age in the late pre-Cambrian period, i.e. 700 million years ago; another in the carboniferous, i.e. 300 million years ago; and a third, known as the pleistocene **ice age**, which began only two million years ago and is probably still going on.

A variety of theories have been put forward about the causes of the latest **ice age**. Some scientists suggested that extra-terrestrial explanations such as variations in the heat radiated by the sun. A recent they suggests that the positions of the drifting continents relative to the pole may account for the earlier ice ages as well as for climatic changes which affected both hemispheres. Ice caps form more easily on land than on oceans, and during the carboniferous ice age extensive ice sheets swept over much of Gondwana land, the massive continent then located over the south pole.

While some scientists have stressed the increase in cold, high elevations on which glaciers could develop during the mountain building activity in the late tertiary period. Others argue that intensive volcanic activity may have ejected great clouds of dust and ash into the upper atmosphere which shielded the earth's surface from the sun's rays. However, critics of both theories have pointed out there have been periods of mountain building and great volcanic activity in the past which were not followed by ice ages. The conflict of opinion and lack of real evidence means that what actually caused the ice ages remains at present a matter of pure speculation.

But most events of the Pleistocene had a profound effect on the landscape and its soils, on the distribution of land and sea and thus on the migration and settlement patterns of our ancestors, e.g. London is built on a series of river terraces formed during the Pleistocene as a result of climatic and sea-level changes. Without them the Thames valley would be an arm of the sea and the Romans would never have chosen the site of London as a crosing place.

All the great harbours of the world such as New York, Vancouver, Sydney and Wellington owe their existence to the melting of the Pleistocene ice caps which raised the sea to its present level. Long before man was settling the land and developing his farming methods, the ice age was moulding it and controlling the ways in which he might use it.

In fact, it would be wrong to think of the ice age as one long freeze-up. In fact, it consisted of a series of alternate cold and warm periods, probably seven cold with six warmer ones separating them. During the cold periods the climate was a good deal colder everywhere than at present, but only during the last three it actually glacial in norhern latitudes, as in Britain these are commonly known as the Anglion, Wolstonian and Devensian stages.

At the above times the polar ice extended for beyond its present limits, in Europe reaching as for south as Southern England, the Netherlands and South Germany. In North America, the ice reached a line running roughly from Seattle in the west to long Island in the east, and passing well south of the Great Lakes. During the last major advance 15,000 to 20,000 years ago, ice covered about 30 percent of the earth.

However, the ice advanced only during the coldest of the cold periods. In the intervening ones, generally called by the name of interglacial periods, the climate was as warm as, or even warmer than the present day. For example, during the last interglacial in Britain, elephants, rhinoceros and hippopotami us wandered as far north as Yorkshire, where there bones are sometimes found in caves or in river and raised beach deposits. So although in Britain the Pleistocene is thought of as an **ice age**, glacial conditions existed for only a very small part of that time. In more southerly countries they never existed at all for what fell as snow in Northern Europe, came down as rain further south. In Africa there is evidence of alternate wet and dry phases, known as *pluvials* and *interpluvials*, which may correspond with the glacial and interglacial periods.

When ice advanced at each time, animals and plants were forced to migrate southwards. When ice retreated, they were able to move north again and recognize the land. Consequently their remains, preserved as fossils by natural burial, can be used to record the climatic changes which the area has undergone. These fossils include vertebrates, such as elephants and rhinoceros for the warm periods, and reindeer and mammoths for the cold.

The distribution of invertebrates such as molluscs can be equally revealing. In a warm period molluscs of southern type migrate northwards, whereas in a cold period they move south again and their place is taken by more boreal species.

It is interesting to note that trees too are good indicators of climate, and the course of an interglacial can often be plotted by tracing the changes in tree cover from arctic tundra to fully developed temperate forest and back again. The evidence is gained by studying the proportions of different kinds of pollen that are preserved, layer-by-layer, in the sediments of ponds or lakes. This has proved to be a most effective way of building up chronological sequences based on climatic change.

The periodic expansion of ice sheets during the Pleistocene inundated above 30 million square kilometres of Northern Europe, North America and Asia, covering large areas with a veneer of glacial sediments. The most characteristic of these is the till, or boulder clay, deposited by the ice itself. Much can be learnt by studying these deposits. It can be seen by an exposure of till on the Yorkshire coast of Eastern England might yield erratics of sharp ige granite, Permian breccia, carboniferous limestones, magnesian limestone and even a battered liassic ammonite. This would tell geologists that the ice which deposited the till came from the lake district in NW Englands by way of the vale of Eden, over the Pennines to Tesside and so down the North Yorkshire coast. Such an assemblage is the "trade-mark" of that particular till; it serves to identify it wherever it is seen, or to distinguish it from other tills which may contain different assemblages.

Almost as extensive as till are the sheets of outwash sand and gravel which spread out from the margin of an ice cap or glacier. The bulk of our present resources of sand and gravel, so vital to the building and heavy construction industries, originated fron this water sorted debris during the Pleistocene. In dry weather a third kind of deposit may become important. Strong winds, blowing over the outwash plains, pick up the silt and carry it in the form of dust storms, perhaps for many miles, before redepositing it as an even spread of loess, blanketing the landscape. Many of the superficial deposits known as *brickearth* on British geological maps are actually thin beds of loess. Much thicker deposits extend from France right across Europe into the CIS. In fact all along the line of the Pleistocene glacial limit.

Impact of the Pleistocene Ice Age

The impact of the Pleistocene ice age on the earth as a whole was not restricted to the deposits of glaciation and a changing pattern of plant and animal life in the immediate vicinity of fluctuating ice-sheets. As the ice was retreating and advancing, so world sea-level falls equally all over the world, leaving raised beaches and river terraces to mark its former height. In the warmer periods most of the icemelts again and sea-level rises, flooding the former land surface and its vegetation, which may later be identifiable as a buried forest.

These world-wide changes in sea-level, brought about by an actual rise or fall of the oceans, are termed eustatic movements. During the coldest part of the last glacial period sea level was at least 100 m lower than now, so the English channel and much of the southern north sea were dry land and other land bridges existed wherever there are now shallow epicontinental seas. On the other hand, if all the existing ice in the world were to melt at once, the sea would rise a further 50 m with catastrophic effect.

While the level of the sea was rising and falling during the Pleistocene, the land in the glaciated areas was rising and falling as well. This is because the continents behave, in the long-term, as though they are "floating" on the denser rock beneath. Extra weight such as thick ice depresses land masses and removal of it allows them to recover their original level. This adjustment, which happens only gradually, is the isostatic effect.

In a glaciated area there will thus be the interaction of two processes, one raising and lowering the surface of the sea. Where the shoreline eventually lies depends on many factors, but broadly speaking in the glacial centres isostasy is more effective than eustasy. In Scandinavia and Northern Scotland, for instance, there is a series of raised beaches representing former shorelines, now elevated well above present sea-level. However, away from the glacial centres, for example in Southern England and France, the eustatic effect is more important, and these areas are characterised by submerged coastlines, drowned valleys and buried forests.

Increasing awareness of the history of the recent past provokes fascinating questions about the future. Will the polar ice caps melt, drowning major population centres.

All that is certain is that the time elapsed since the end of the last glacial period is less than the duration of most of the intergalcials, so there is no real reason to assume that the ice age is over and done with.

Perhaps post-glacial time is just another interglacial period. What is more, perhaps the mid-point of that interglacial is already past, for there was undoubtedly a time, around 4000 BC, when the climate was shightly warmer than it is now. One consoling thought is that climatic change takes place very slowly. It has to take much more time to change the climate, so we need not to worry about another glaciation. But present global warning is serious problem that enhance great climate change.

21 | Work of Sea

Our coasts are always assaulted by seas. The pounding action of the waves is not only destructive. Cliffs and other coastal landforms are being eroded, the resulting rock debris may be deposited by that same wave action elsewhere.

POWERFUL WAVES

Most coastal changes occur under storm conditions, when the destructive powers of waves are at their graetest. The energy of a wave depends on its length, its height and its swiftness or celerity. Variation in any one of these characteristics will change the ability of a wave breaking on a coast to erode and move material.

However, the most important factor in the power of a wave is its height. Because this depends in part on wind speeds, the highest waves most commonly occur in storms, when Gales whip the sea into a furious assault on the coast. The crash of storm waves on a cliff traps pockets of air in the rock cavities and compresses them. Then as waves fall back, the air expands explosively, throwing spray, pebbles and shattered rock high into the air. Further erosion is the result of the corrosive action of the debris hurled by waves against the coasts. The material itself is worn into smaller particles by the constant grinding. The immense amount of damage that can be caused by storm waves was dramatically illustrated in January 1953, along the coasts of the north sea. Under strong northerly winds and the high tide, a surge of water was forced into the souhern part of the north sea. The effect was devastating. In many places along the eastern coast of England, the beach was completely washed away by the sea, and once this protection was lost, the cliffs were exposed to rapid erosion. In some areas of Lincolnshire, low cliffs were cut back by more than 10 metres.

The sea can also change a shoreline considerably without actually eroding the cliffs or beach. On tropical coasts, huge waves are generated by violent hurricanes. In 1960, hurricane donna in two days shifted an estimated 176,500,000 cubic feet, of sand from one part of the Florida coast to another. Under normal conditions it would take about 100 years to transport that amount. Although very little erosion had occured, at the end of the storm, each resort had an entirely new beach similar in all respects to the pre-storm beach.

In fact, the clearest example of the erosive action of the sea is the way a cliff is undercut by wave action, and then eventually recedes as the unstable slope above collapses. Cliffs are undoubtedly the most striking landform to be seen along a coast, and although their height is entirely determined by the relief of the land, the sea can have dramatic effects. In England, for example, some parts of the Isle of sheppy are retreating by more than three metres every year. On another part of the eastern coast of England, the cliffs are being eroded even further as much as 10 metres a year.

It is important to mention here that the sea can attack the cliffs only within a very restricted vertical range, in effect up to a level reached by the highest waves. Concentrated at the base of cliff, the sea's erosive force obviously varies with the strength of the waves. In sheltered locations for example, cliffs are eroded much more slowly than those on exposed coasts. However, the other major factor affecting the rate of erosion is the geology of the area. Cliffs composed of soft rocks, such as clays or glacial sands, can be attacked and the rock debris washed away very quickly.

Cliffs of granite offer much greater resistance to the pounding force of the sea, even along exposed shores such as land's end at the tip of the South-West England and the stromy Cape Horn of South America. Where harder rocks alternate with softer ones, the sea often carves out bays and coves in the less resistant rock leaving the harder ridges jutting into the sea as headlands. Straight shorelines are characteristic only of faulted coasts and those formed of rocks of generally uniform resistance to erosion. Along the English channel for example, the famous white cliffs of Dover and the seven sisters coast near Eastbourne are composed of chalk of very even texture.

Marine erosion at the cliff-base occasionally creates unusual landforms. Most hard rocks have fault joints and other lines of weakness which are exploited by the sea, sometimes cutting inlets and caves deep into the cliffs. The explosive pressure of trapped air and surging water inside a cave may be sufficient to erode upwards through the roof to form an opening known as a blowhole from which clouds of spray may shoot wards. Quite often, caves hollowed on both sides of a headland join up to from a natural arc. In time, when the top of the arc collapses, the remnant of the headland stands as a detached pillar, known as a *stack*. All tall, offshore rock pinnacles are called stacks, irrespective of how they were formed. Many well-known examples occur around the coasts of Britain, such as the chalk pinnacles known as the needles off the Isle of Wight, and the old man of hoy, a 137 m high pillar of old red sandstone, in the Orkney Islands of Scotland.

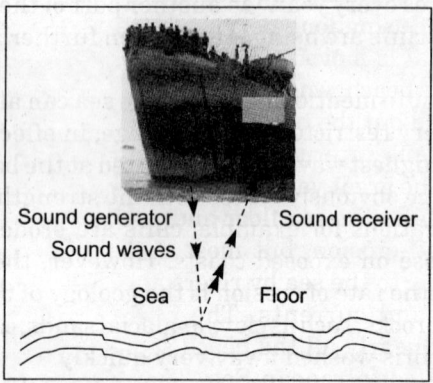

Fig. 21.1: *Determination of the floor of seas by echo method.*

22 Working Stages of Beaches

A beach is a deceptively transient feature a sloping accumulation of loose material which may consist of boulders, shingle, pebbles, sand, mud and shells along the sea-shore. Movement of material both up and down and along the beach ensures that is constantly changing. Beaches may be removed overnight by a violent storm, as occurred in 1953, along the Lincolnshire coast but they usually build-up again during long period of calm weather. Often a thin layer of material is moved nearly continuously, being deposited at one end of the beach, washed by longshore drift along the length of it, and carried out the other end.

The continuous interplay of wave sand this loose material contributes to the great variety of beaches. Sand and shingle may be washed up from the sea floor and a small amount of debris is provided by cliff-erosion, but most of the material comes from sediment carried to the sea by rivers and then transported along by the waves and currents. The river Nile, for example, is responsible for nearly all the beach sediments in the south-east cornor of the Mediterranean Sea. The remains of shells, corals and other organisms may be a further source of beach material.

It is often possible to trace the varied origins of material. For example, at cap Griz Nez, near the French channel port of Calais, the beach consists of huge boulders more than a metre across, which have fallen from thick limestone lands in the cliff face. Fillint cobbles the size of a frist are found on Chesil beach in Dorset, having been eroded from the chalk on the channel floor, then washed up by the waves.

Coarse sand is common in the beaches of cornwall, having been eroded from the granite rock inland, while extremely fine sands from the old and new red sandstone areas provide a popular holiday beach at Weston super Mare on the Bristol channel coast.

On all beaches, eroded material is gradually broken down into smaller pieces by the constant pounding of the sea. The surge or swash of waves pushes pebbles and sand up the beach, while the backwash or underwater currents drag the material down the slope. Generally, the coarser the material, the steeper the beach slope will be. The steep profile of the shingle beach at Chesil beach, for example, is partly due to the large size of the flints, whereas the beach at Weston super Mare slopes gently because of its fine sand. Where a mixture of sediment size occurs, as at many beaches of East Anglia, the coarser material tends to gather at the highest part and there is usually a marked change in slope between this and the finer sediments towards the low water level.

The shape of a beach is also affected by variations in the movement of waves. Once the direction of prevailing waves or the swash and backwash change, the beach is modified, either by erosion or deposition. If the beach is battered by high waves, as often occurs in winter, the loose material racked from the beach is carried back to the sea, producing a more gentle slope. By contrast, when these steep waves are replaced by more gentle swell waves material tends to be built-up and the beach slope is increased. This produces the generally steeper, larger beaches of summer.

Waves tend to approach the shore at right angles. This is true even of a cape and bay coastline, for as the waves approach the shore, they first encounter the shallower water opposite the headlands and are slowed down. Opposite the bays, however, the water usually remains deep, so that the waves advance more rapidly. As they too begin to "feel bottom", they slow and swinging effect, known as *wave-refraction*, occurs by which the line of advance of the waves becomes generally parallel to the shore so that they break head-on.

In many places, however, waves are not fully refracted, and approach at an angle to the coast. This results in longshore drift the movement of shingle and sand along a coastline. Where longshore drift is strong, an abrupt change in the direction of the coast or the entry of a river can produce a spit, a narrow ridge of sand, gravel and pebbles piled up by the waves.

Orford Ness on the Suffolk coast is a shingle spit which first sealed off the estuary of the river Alde and then extended southwards for

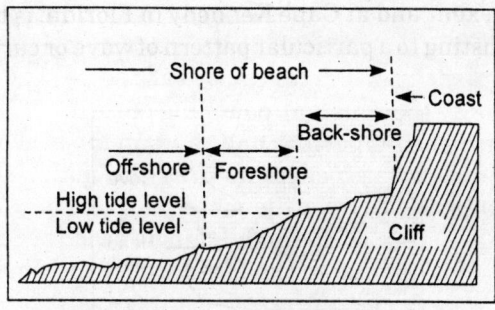

Fig. 22.1: *Coast and shore and sub-division.*

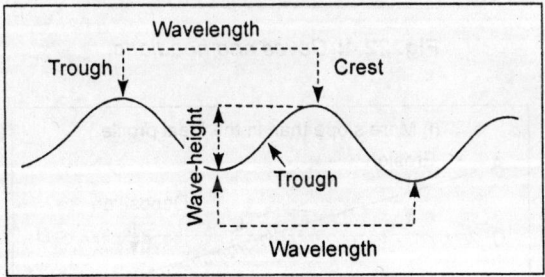

Fig. 22.2: *Wave.*

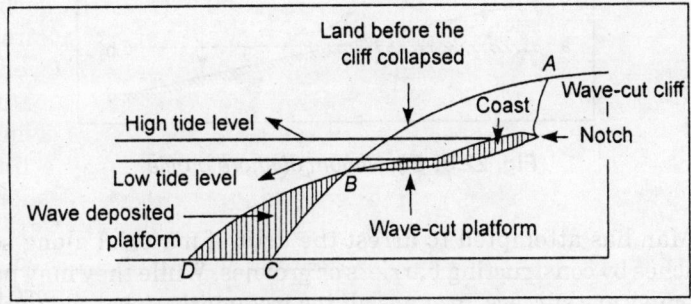

Fig. 22.3: *Cliff and wave-cut platform.*

several kilometres across the mouth of the river Butley. The town of Orford lies inland between the two rivers, but 750 years ago it was a port facing the open sea. The curiously shaped triangular beach known as a *cuspate foreland*, and found for example, at dungeness in kent and at Cape Kennedy in Florida, is the result of material adjusting to a particular pattern of wave or current action.

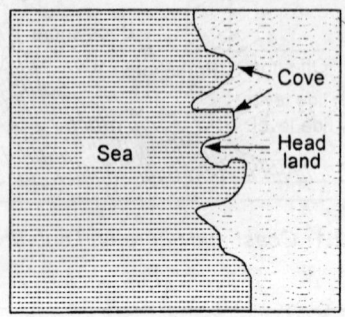

Fig. 22.4: *Coves and headland.*

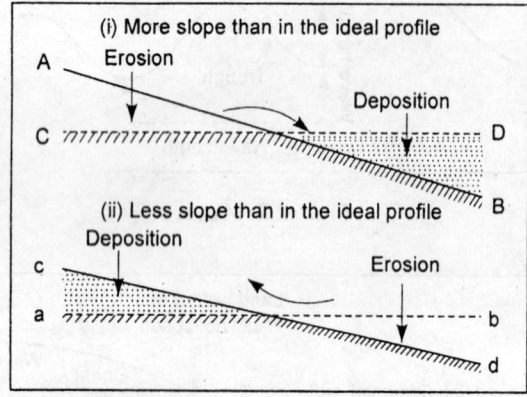

Fig. 22.5: *Equilibrium of shore profile.*

Man has attempted to arrest the drift of material along some beaches by constructing barriers or groynes. While they may prove a temporary success, groynes all too frequently upset the natural relationship between beach supply, wave action and sediment transport, so that at the end of the protected zone serious erosion occurs. As an alternative to this a number of attempts have been made to artificially replenish beaches by dumping material of an appropriate size. For example, Portobello beach near Edinburgh, robbed of its sand by a long history of erosion, has received this

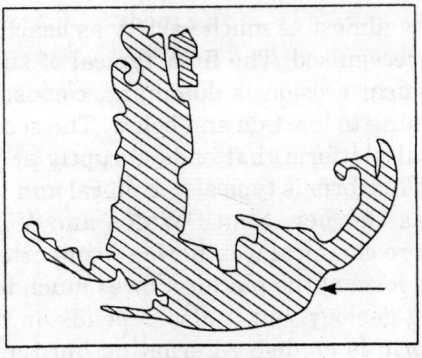

Fig. 22.6: *Hook.*

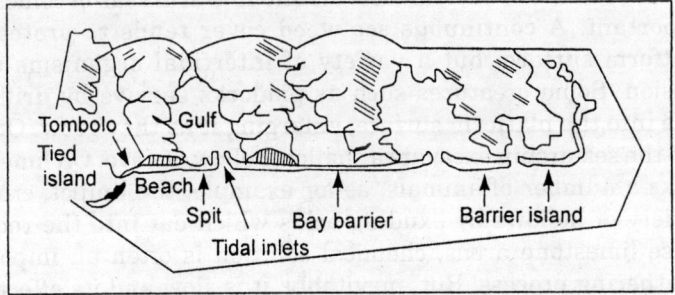

Fig. 22.7: *Sea waves and landforms made by them.*

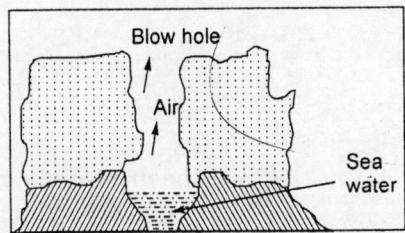

Fig. 22.8: *Blow holes or spouting horns.*

treatment with some success and the technique has been quite widely adopted in California and Florida.

It is interesting to mention that along rocky coasts, a level rock shelf or shore platform extends seawards from the cliff base. Shore platforms show almost as much variety as beaches but two basic forms can be recognised. The first, typical of storm wave coasts where mechanical erosion is dominant, consists of an inclined surface stretching to low tide and below. The second consists of a near horizontal platform that ends abruptly at what is called a low tide cliff. This form is typical of tropical and warm temperate areas, such as the new South Wales and Victoria coasts of Australia, where chemical weathering is important.

The surface form of the platform owes much to the waves and the underlying geology, but it also depends on the processes by which a platform is eroded. Alternating hard and soft sands in shore platforms are eroded at different rates so that miniature scarps and values are produced. Biological activity is often a notable feature of shore platforms and in warm seas becomes very important. A continuous sea-weed cover tends to protect the platform surface, but a variety of intertidal organisms cause erosion. Some creatures such as piddocks survive by drilling a hole into the platform surface, enlarging it as they grow. Others, like the sea urchin excavate a shallow hollow or cave. On limestone rocks a number of animals, as for example, the limpet, create a variety of hollows by exuding acids which eat into the rock. In these limestone areas, chemical solution is often an important weathering process. But, inevitably, it is slow and its effect may be overtaken by other, more powerful, erosive processes such as the pounding of storm waves.

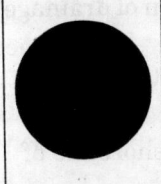

Geomorphology: Objective Style

1. A special type of sinkhole in karst region with steep sides and flat floor is
   - (a) Blind valley
   - (b) Poljes
   - (c) Lapies
   - (d) Bolson
   - (e) Uvalas

2. A general term for sinkholes in a karst region formed due to solution action is
   - (a) Polje
   - (b) Karst window
   - (c) Solution pan
   - (d) Dolines
   - (e) Both polje and dolines

3. The synonymous term for window is
   - (a) Nappe
   - (b) Cockpit
   - (c) Clippe
   - (d) Fenster

4. A highland between two streams belonging to the same drainage pattern is the
   - (a) Tomolo
   - (b) Inter-flux
   - (c) Spits
   - (d) Fletiorn

5. Ria coast consists of broad valley and their sides have gentle slopes. They are the effect of
   - (a) River
   - (b) Glaciers
   - (c) Winds
   - (d) Oceanic waves
   - (e) Both wind and glaciers

6. The term "lost river" has been applied to streams which disappeared completely underground in a
   - (a) Granitic terrain
   - (b) Metamorphic terrain
   - (c) Dolomitic terrain
   - (d) Limestone terrain

7. When the stream valleys are guided by the fold they are parallel to the system of fold axes, this pattern of drainage is called
   - (a) Dendritic
   - (b) Radial
   - (c) Trellis
   - (d) Annular
   - (e) Barbed

8. Bays and headlands are generally found in shoreline of
   - (a) Submergence
   - (b) Emergence
   - (c) Neutral type
   - (d) Faulted type

9. Chemical weathering is more effective than mechanical weathering in
   - (a) Semi-arid region
   - (b) Arid region
   - (c) Wet or humid region
   - (d) Both semi-arid and humid regions
   - (e) All of the above

10. The temperature of freshly erupting lava ranges between
    - (a) 900–1200°C
    - (b) 700–900°C
    - (c) 1200–1300°C
    - (d) 800–1400°C

11. Name the highest active volcano in the world
    - (a) Etna (3,300 m)
    - (b) Cotapaxi (5974 m)
    - (c) Sangay (5980 m)
    - (d) Fuego (4786 m)

12. The second largest island in the world, i.e. New Guinea, is located the
    - (a) Atlantic ocean
    - (b) Pacific ocean
    - (c) Indian ocean
    - (d) Bay of Bengal
    - (e) Arabian sea

13. The gas found dissolved in the sea in an amount greater than its amount in the atmosphere is
    - (a) Nitrogen
    - (b) Oxygen
    - (c) Argon
    - (d) Carbon dioxide
    - (e) Methane

14. Fumeroles emitting sulphureous vapour, CO_2 and boric acid vapour are respectively known as
    - (a) Moffette, Soffioni and Salfatarous
    - (b) Soffioni, Salfataras and Moffettes
    - (c) Salfataras, Moffettes and Soffioni
    - (d) Salfatarous, Soffioni and Moffettes

15. At what rate the delta of Ganga is growing
    (a) 2 feet/year   (b) 3 feet/year
    (c) 5 feet/year   (d) 10 feet/year
    (e) Growth is insignificant

16. The Laccadives and the Malidives, occuring within the Arabian Sea, furnishes a beautiful example of
    (a) Fringing reefs   (b) Coral Islands
    (c) Shingle beach   (d) Atolls

17. Which of the following feature is formed below the base level of erosion.
    (a) Bolsons   (b) Stacks
    (c) Monodnock   (d) Fjords
    (e) Kettle hole

18. Alluvial fan coalesce it generally gives rise to
    (a) Flood plain   (b) Delta land
    (c) Ox-bow lake   (d) Feature named Bajada
    (e) Hammadas

19. The name given to the mountain ranging beween Indus river in Kashmir and the Pamir Plateau is
    (a) Purvanchal   (b) Himalayan
    (c) Trains   (d) Karakoram

20. How the bottom topography of an ocean are shown
    (a) Hypsographic curve   (b) Contour maps
    (c) Hydrostatic curve   (d) Hyther graph
    (e) Choropleth map

21. Match the following, a, b, c, d, e with corresponding 1, 2, 3, 4, 5, 6.
    (a) Arid land equivalent of a   (1) Badlands
        Peneplain
    (b) Wind blown glacian silt and   (2) Erratic
        clay
    (c) Arid climate terrain in   (3) Pediment
        closely dissected soft rocks
    (d) Glacial boulder resting on   (4) Loess
        foreign rock
    (e) Principal of continuity of   (5) Barchean
        geologic process   (6) Uniformitarianism

22. The earth's most stable environment is found in/on
    (a) High mountains   (b) Deep-sea floor
    (c) Inosphere   (d) Polar regions
    (e) Semi-arid regions

23. An example of block mountain is
    (a) Himalaya   (b) Andes   (c) Alps
    (d) Vosges   (e) Appalachain

24. Iceburgs of masses of glaciers with 1/9th of their part
    (a) Above sea level   (b) Below sea level
    (c) Equal to sea level (d) Independent to sea level

25. Name the dry land extending below sea level
    (a) Death valley   —   California (86 depth)
    (b) Dead sea   —   (390 m depth)
    (c) Caspian sea   —   (946 m depth)
    (d) Both (a) and (b) are correct

26. Nearly the whole land surface of earth was covered by great sheet of ice during
    (a) Cambrian   (b) Precambrian
    (c) Cretaceous   (d) Pleistocene
    (e) Jurassic

27. The well-known dead sea is found to be located in a
    (a) Rift valley   (b) Canyon   (c) Intermont plain
    (d) Between the chains of mountains

28. The attitude of the snow line varies to the latitude of the place concerned.
    (a) Directly   (b) Inversely
    (c) Logrithmically   (d) Do not depend upon it

29. Exfoliation is accomplished when the surface layer of rock pulls off because of.
    (a) Chemical change   (b) Excesive heating
    (c) Release in pressure after the overlying layers are removed.
    (d) Leaching and bacterial action.

30. Formation of travertine terraces are due to
    (a) Wind action   (b) Fluvio-glacial action
    (c) Ground water action (d) Glacial action

31. Tropical desert all over the world are generally located only on the
    (a) Western margins of the continents
    (b) Eastern margins of the continents
    (c) South Central part of the continents
    (d) Northern margin of the continents
    (e) Found anywhere in the cotinents

32. Deposits of weathered material found at or near its source are termed as
    (a) Collovial deposits   (b) Eluvial deposits
    (c) Alluvial deposits   (d) Placer deposits
    (e) Residual deposits

33. A spit is a bar of sand or shingle.
    (a) Which is in offshore and runs parallel to the coast
    (b) Which joins and island to the main lands
    (c) Whose one end is connected with land and the other runs into the sea.
    (d) Which runs across a bay and connects two headlands.

34. The uprush to tidal waves of certain height through the river mouth is called.
    (a) Seiches   (b) Tsunamis   (c) Spring tide
    (d) Tidal bore   (e) Neap tide.

35. The cycle of erosion are at many times interrupted, such a type of interruption is
    (a) Retrogression   (b) Aggradation
    (c) Degradation   (d) Rejuvenation

36. Run-off and erosion would probably be greatest on a land area that is
    (a) Flat lying and slightly covered with vegetation
    (b) Sloping and barren or devoid of any vegetation
    (c) Gently sloping and covered with grass
    (d) Sloping and contaur ploughed.

37. What are "alkali flats"
    (a) Salt lake of Utah
    (b) Deposited salts in the coast of Atlantic ocean
    (c) Lake of fresh water near the sea-coast
    (d) Thin deposits of alkali material at bottom of sea

38. The division of the world into the major natural regions is primarily based on

    (a) Rock types present
    (b) Rainfalls
    (c) Annual temperature ranges
    (d) Types of climate
    (e) Soil types

39. Which of the following minerals does not show alternation

    (a) Biotite   (b) Olivine
    (c) Othoclase   (d) Quartz

40. One method of showing the rate of erosion on a steep hill is to

    (a) Plant trees on hill slope
    (b) Remove rocks from the hill side
    (c) Make the slope gentle by cutting
    (d) Dig a drainage ditch at the bottom of the hill.

41. When the eruption of volcano takes place the form of explosion of tremendous force, which ejects numerous fragments of different size but without lava, the volcanoes are of

    (a) Volcanoes type   (b) Barindaision type
    (c) Hawanian type   (d) Vesuvian type

42. The size particle in case of suspension is

    (a) 10^{-5} cm   (b) 10^{-8} cm
    (c) 10^{-6} cm   (d) 10^{-4} cm

43. What is rift valley

    (a) It is a deep valley formed in between the mountains
    (b) It is a valley on the sides of which are huge mountains
    (c) It is a subsidized land leaving a long and narrow opening
    (d) None of the statement suits for rift valley

44. A plain formed on the both side of rivers by the deposition of sediments carried by it is

    (a) An alluvial fan   (b) An eskar
    (c) A flood plain   (d) A delta
    (e) Levees or revees

45. The famous American "dust bowl" has been created by
    - (a) Soil erosion
    - (b) Land sliding
    - (c) Soil conservation
    - (d) Glacial deposition
    - (e) Meandering of rivers

46. Cosmic rays
    - (a) Come from outer space
    - (b) Come from the sun
    - (c) Is another name of corenkov radiations
    - (d) Are having unknown wavelengths

47. Choose among the following the largest island in the world
    - (a) Greenland
    - (b) New Guinea
    - (c) Andaman Nicobar
    - (d) Madagaskar
    - (e) United Kingdom

48. Solar radiations, which are having visible and near infrared wavelength supplies how much energy to the geomorphic system.
    - (a) 40.34%
    - (b) 58.39%
    - (c) 89.47%
    - (d) 92.96%
    - (e) 99.98%

49. The usual altitude of plains are
    - (a) Sea level or below it
    - (b) Less than 300 metres above sea level
    - (c) Greater than 660 metres above sea level
    - (d) Less than 500 and greater than 400 metres from sea level.

50. Solar radiations provides energy for geomorphic changes at a rate of about—faster than all other energy sources
    - (a) 200 times
    - (b) 350 times
    - (c) 2549 times
    - (d) 4000 times
    - (e) Rate is not clearly known

51. On what basis the continents are divided into physiographic Provinces
    - (a) On the drainage of major rivers
    - (b) Climate and vegetation
    - (c) The type of bed rock
    - (d) Distinctive landscape pattern

52. The statement "Landscape is a function of structure, process and stage" is given by
    - (a) J-Hutton
    - (b) Penk
    - (c) Davis
    - (d) Plafair

53. Circulation of water through hydrosphere, atmosphere and lithosphere continuously is known as
    (a) Evapotransformation   (b) Water circulation
    (c) Hydroequilibrium cycle   (d) Hydrological cycle
    (e) Hydrostatical equilibrium

54. In karst cycle the beginning of diversion of surface drainage to sub-terranean routes with dolines as the characteristic landform is a feature of
    (a) Youth stage   (b) Mature stage
    (c) Old stage   (d) May form at any stage

55. The annual alternation of torrential rain and prolonged drought create a distinctive landscape known as
    (a) Badland   (b) Wadi   (c) Savanna   (d) Shingle

56. A common term for the depositional features in a limestone cavern which develop vertically, horizontally, obliquely, incurved, upward and downward direction is
    (a) Stalactite   (b) Flowstone
    (c) Helictite   (d) Stalagmite

57. What is a tectonic plateau
    (a) A plateau produced by uplift of a portion of earth's crust
    (b) A plateau formed on account of extensive sheets of lava getting piled up one over the other
    (c) A plateau associated with the formation of fold mountains
    (d) None of the above

58. Most characteristic drainage pattern of trap area is
    (a) Dendritic   (b) Barbed   (c) Rectangular
    (d) Parallel   (e) Annular

59. Ox-bow lakes are generally associated with the—stage of a stream
    (a) Youth   (b) Mature
    (c) Old   (d) In any stage of a stream

60. Residual clayey soil found as mantle covering in Karst region is known as:
    (a) Lapies   (b) Terra Rossa   (c) Oxisol
    (d) Sinters   (e) Pedosol

—————————— ANSWERS ——————————

1. (b)	2. (d)	3. (d)	4. (b)	5. (a)
6. (d)	7. (c)	8. (b)	9. (e)	10. (a)
11. (b)	12. (b)	13. (d)	14. (c)	15. (a)
16. (d)	17. (d)	18. (a)	19. (a)	20 (a)
21. (a) – 3,	(b) – 4,	(c) – 1,	(d) – 2	(e) – 6
22. (b)	23. (d)	24 (a)	25. (d)	26. (d)
27. (a)	28. (b)	29. (c)	30. (c)	31. (a)
32. (b)	33. (c)	34. (d)	35. (d)	36. (c)
37. (b)	38. (d)	39. (d)	40. (a)	41. (b)
42. (d)	43. (c)	44. (c)	45. (a)	46. (a)
47. (a)	48. (e)	49. (b)	50. (d)	51. (d)
52. (b)	53. (d)	54. (a)	55. (c)	56. (c)
57. (a)	58. (b)	59. (c)	60. (b)	

Part II

HYDROGEOLOGY

HYDROGEOLOGY

Geology is the study of the earth as a whole. If anybody want to know everything about our mother earth he has to study geology. It includes many branches of science.

Hydrogeology is an important branch of geology. It mainly deals with the water on the earth, above the earth, and underneath the earth. One can easily understand that *hydro* means water, *geo* means earth and *logy* means study.

Thus, hydrogeology is that branch of geology which deals with all states of water of the earth as a whole. A hydrogeologist is a scientist, who deals with all the activities of water of the earth. A geologist at drilling site is called as a "well-site geologist".

23 | The Global Water Budget

The globe earth having the surface area of which 29% is occupied by the land and the remaining 71% is covered by oceans. These oceans with a mean depth of 3.8 km hold as large as 97% of the earth's water, while 2% is frozen in ice caps. The deep ground water accounts for 0.31%. Thus, 99.31% of water on earth is of no healthy use to man. It is fact that at any instant, rivers and lakes hold only 3% of fresh water (or 0.00095% of total water). The global annual precipitation and evaporation are estimated to be equal to about 100 cm each. Therefore, the global average annual precipitation volume is about 510100 km³. The average moisture available in the atmosphere at any time is only 12900 km³. This implies that the entire atmospheric moisture must be replaced 40 times each year. In other way the average residence of the atmospheric moisture is slightly more than 9 days.

Distribution of World's Water Resources

Location	Water volume in km³	Percentage of total water
Fresh water lakes	125000	0.009
Saline lakes	104200	0.008
Stream channels	1200	0.0001
Ground water (< 0.8 km deep)	4168200	0.31
Ground water (> 0.8 km deep)	4168200	0.31

Distribution of World's Water Resources *(Contd.)*

Location	Water volume in km³	Percentage of total water
Soil moisture, etc.	66700	0.0005
Ice caps and glaciers	29177300	2.15
Atmosphere	12900	0.001
Oceans	1321310000	97.2
Approx. Total	13.6×10^8	100

24 Practical Applications of Hydrogeology

As hydrogeology has many practical applications some of them can be discussed which are as below.

STRUCTURAL DESIGN

Structures like spillway, dam, current, highway bridge, rail bridge, etc. having design which may be considered to consist of mainly three parts, viz. hydraulic design, hydrologic design and structural design. Hydrologic design deals with the estimation of quantities of water to be handled at the site of the structure; specifically their time distribution, time of occurrence and frequency of occurrence. Structural design ensures the stability of the chosen section against the water pressure and other pressures. Hydraulic design provides the best suitable section of the structure to cope with the waters estimated. So hydrology plays an important role in the design of any structure. We may see it as, if we consider the design of rail or highway bridges, or culverts, their cost adds up to considerable amount in the total cost of transpertation system.

In fact, it is not possible to design all of them to with stand serve most and rare floods. Hence, the problem is to find not only the magnitude of flood discharge but also probability of occurrence so that the structure may be designed for such a discharge at which the additional cost of protection against higher flood discharges may be balanced against the possible cost of replacing the structure.

Hydrology provides basic data for such studies which come to the design of spillway the hydrologist must evaluate not only the probability and magnitude of design flood but also the effect of reservoir behind the spillway on the distribution of the design flood. It the magnitude of the design flood is 2000 m^3/s, because of the storage effects of the reservoir. This problem is to be called as *flood routing*. Similarly, the hydrology of urban areas play an important role in the efficient and economic design of drainage structures such as storm-sewers.

IRRIGATION

The essential part of the design is to irrigate the ayacut under the project or to meet the demands of other purposes from time to time and for this provision of adequate storage facilities at irrigation projects and other multipurpose projects are made. In arriving at the storage capacity of the reservoir, the evaporation, seepage and other losses must be properly accounted for. Till recent times the practice has been to design the project in isolation with the available data however meagre it may be. The integrated approach in designing a system as a whole has been proved beneficial. Large river basin is taken as a system and all the projects in the system are together optimally designed and operated to achieve the stipulated objectives. Hydrogeology of the basin provide the required information regarding the bed rock and stream flows at various project sites, their variability and frequency distribution. This information is required in arriving at the capacities of the reservoirs and in deciding the operation policies. Scientists form geology side may also be called upon to help while studying the alternatives such as artificial recharge for lift irrigation scheme.

MUNICIPAL AND INDUSTRIAL WATER SUPPLY

In locating the important industries availability of water is often the most important factor has considerable effect on the growth of municipalities. Hydrologist generally give answer about the flow in the nearby stream in whether it is sufficient to meet the needs of municipal city or industry. Hydrogeologist also suggest, whether the ground water in that region can be tapped economically to meet the above needs without over exploitation or whether stream flow and ground water can be utilized conjunctively, because municipality and industries are backbone of human settlement.

HYDROPOWER

Hydrologic studies are essential for the development of water power planning. The feasibility of run-off river plant operating with pondage can be determined by a reliable prediction, which is needed of the absolute minimum daily flow that may be expected in the stream and of the percentages of time that various other flows may be expected to exist. That means a reliable flow duration curve is required. The absolute minimum flow decides the prime capacity (firm power) of the plant, while the additional flow data is useful in estimating the amount of power that will have to be obtained from other sources for any increase in the installed capacity. For storativity of water for storage plants. However, low seasonal flows rater than low daily flows are important. Scientists are to give the reliable hydrologic predictions of stream behaviours so that whether the reservoir can be emptied boldly permitting the shutdown of the other thermal resulting in fuel saving, or should it be drawn on sparingly so that shore will be ample water for future dry periods.

NAVIGATION

Navigation projects require water sufficient of lock gates and maintain minimum draft, wherever this water can be obtained and preserved and the effect of navigation structure on stages in the river especially during floods, etc. are the problems in navigation faced by scientists from geology side. If we observe the streams loaded with heavy silt and clay, this creates complex problems.

EROSION AND SEDIMENT CONTROL

As we know that excessive erosion in the catchment feeds sediment into the runoff carried by the stream which leads to many undesirable effects. The reservoirs got reduced their economic life span drastically due to lose their capacity at faster rate. Due to excessive erosion and sedimentation, tons and tons of fertile top soil got lost every year which results in reduction of crop yield. The problem of erosion control is mainly linked with the phenomena of overland flow and infiltration. Hydrology of the catchment + knowledge of the existing watershed management practices help in finding out effective erosion control measures suitable for the given soil conditions. Effective erosion control measures include fixing of judicious crop pattern and cropping procedures, formation of contour bands, afforestation, etc. Effective erosion control measures decrease the sediment load in the stream and also reduce the flood discharges due to increased infiltration opportunities in the catchment area.

FLOOD CONTROL

The most commoly used flood control structures are reservoirs, levees, channel improvement and channel diversion. The technique of flood routing is essential to intelligent and economical planning of flood control projects. As we know that flood control problems are complicated because any type of flood control project modifies the natural regime of the stream and thus in the process of protecting one area, it may increase the flood damage in another. Flood control measures are generally effective when they are associated with flood forecastings. The joint work of hydrologists and meteorologists utilised in planning the advanced evacuation of threatened areas. Flood plain zoning is also considered as one of the flood control measures, when adequate hydrometeorological information is available.

POLLUTION ABATEMENT

As a stream is a natural water purification system, it is generally considered permissible to allow disposal of certain amount of sewage into streams. The purification in the stream is a result of bacterial action and aeration. But indiscriminate disposal of sewage from cities and industries into the nearby streams, which has been the general tendency, results in health hazards to the public and destruction of the fish and other wild life. Nature of stream flow and its variability as studied by a hydrogeologist helps sanitary engineer regarding the level to which the sewage has to be treated before it is disposed of into the stream and also the quantity of sewage that can be allowed into the river without causing any damage to the downstream regions. The augmentation of stream flow by way of a storage dam upstream is good to keep the sewage concentration within allowable limits and it more economical than additional sewage treatment plants.

HISTORICAL DEVELOPMENT

Date back to prehistoric times, we find evidence of utilisation of water resources like that of wells and early irrigation system. We are unable to know that who perceived the concept of hydro-logic cycle first as we know it today. The stream flow's appearance in the river even long after the cessation of rainfall was rather a puzzle for pre-historic men. Early philosophers like Plato and Aristotle, gave speculation which led to the erroneous concepts such as subterranean condensation and underground sea which according to them were responsible for the stream flow even during dry periods. Vitruvius gave much like our modern concepts, who postulated that ground water is derived only from rain and snow. The first to advocate the infiltration theory were Leonardo da Vinci (1452–1519) and Bernard Palissy (1509-1589) to explain the occurrence of water in springs and dry weather flow in rivers.

The quantitative measurement of hydrologic phenomenon began only in the 17th century. According to Pierre Perrault, the estimated stream flow of Seine river was approximately only on sixth of the measured precipitation volume on the basin. Another scientist, viz. Mariotte (1620-1684), made detailed velocity measurements using floats and made computations of discharge and established the findings of perrault on the basis of more exact run of data. On the basis of experiments on evaporation of salt water and Halley (1620-1687) established that evaporation from the sea is adequate to supply all the water that the rivers discharge into the sea. That is why Meinzer considered Perrault, Mariotte and Halley are to be considered the founders of hydrology, while Vitruvius, Vinci and Palissy were discoverers who advocated correct hydrologic principles.

There were the significant contributions to hydrology in the 18th century by Pitot, who made contribution in the development of velocity tube and Chezy gave derivation of flow formula in channels. In 19th century, the notable developments regarding surface water as weir formula by Francis, venturimeter by Herschel, Ganguillet and Kutter's formula to determine Chazy's C while Darcy, Dupuit, A Theim, G. Thiem, Forchheimer, etc. made in ground water development.

Modern hydrogeological methods mainly developed in the 20th century with the unit hydrograph concept as praised by Sherman and Hazen used statistical methods in hydrologic studies. In the later part of the 20th century, activity in the hydrologic study and research got momentum. In recent years use of systems approach, development of mathematical models, digital stimulation, application of stochastic and probabilistic theories to hydrologic phenomena, etc. are some of the significant developments as we know today. UNESCO, in 1965, organised the "International Hydrologic Decade" to focus the attention of developing countries on importance of hydrology but in the year 1975, it was changed to "International Hydrologic Programme".

25 Water

Two parts of hydrogen and one part of oxygen make water, as water is the mother of life. If water will be pure the life will be pure. Water shows the past, present and future conditions of human settlements.

Hydrology is the science which deals with the movement of the water on the ground, under the ground, evaporation from land and water and transpiration from the vegetation. Not only this is included in it but also the way of going back the water to the atmosphere from where it precipitates.

The water in atmosphere due to evaporations and transpiration again comes back in the form of precipitation under favourable climatic conditions.

Figure 25.1 shows that the hydrologic cycle of water made by the process as due to sun's heat the water from the surface of the earth, lakes, rivers, seas, etc. evaporates and rises upwards. At high attitude due to reduction in the atmospheric pressure these water vapours expands by also or being energy from the surrounding air, which cools down. The atmospheric capacity at high attitude depend on its temperature and moisture content. When it falls below the dew point, it can hold up the excessive moisture, which starts falling in the from of rain, hails, dew, sleet, frost and precipitation. When the temperature falls too much it may cause snowfall which is indeed a form of rainfall. Various factors such as temperature, atmospheric pressure, velocity of wind, height of mountains in the region, presence of forests, position of land and water areas, etc. and their complex relations are responsible for the precipitation.

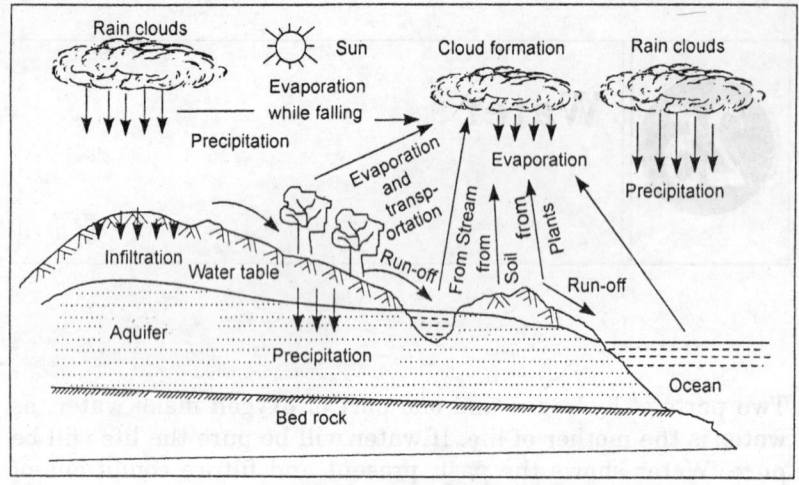

Fig. 25.1: *Descriptive representation of hydrologic cycle.*

Precipitation is the primary source of all water whatever available on the earth and includes rainfall, snowfall, hail and sleet. All the sources of water receive their supply of water from the precipitation. The water of precipitation further goes in the following ways:

(a) *Run-off:* A portion of precipitated water flows over the ground in the form of rivers and streams and some water flows towards lakes and ponds and is collected there.

(b) *Percolation:* A portion of precipitated water percolates in the ground and is stored there in the form of sub-soil or ground water.

(c) *Evaporation:* Some portion of the precipitated water is also evaporated from the lakes, rivers, reservoirs and wet surfaces in the forms of vapour due to the heat of the sun. The portion so evaporated is called *evaporation.*

(d) *Transpiration:* The roots of the trees which usually suck water from the ground and some portion of it evaporates in the atmosphere through leaves in the form of transpiration.

Details of whole play role of water above the earth's surface, on the earth's surface and beneath the ground, can be known by latter chapters of this book.

# 26 | Water Above Earth's Surface

Water above earth's surface is generally present in the form of precipitation. As there are four basic conditions which are required for the occurrence of precipitation, viz.

(a) Accumulation of moisture of sufficient intensity to account for the observed rates of precipitation.
(b) Cooling of air to the dew point temperature to produce saturation condition.
(c) Condensation.
(d) Growth of small water droplets to precipitation size.

It is clear that the simple mass conservation requires that an equality between evaporation and precipitation be maintained. In fact, the observed evaporation rates support this. So it is by evaporation that sufficient moisture accumulates in atmosphere. The other three conditions have been discussed below.

Cooling

Cooling is of three types, *viz.* (a) cyclonic cooling, (b) orographic cooling, and (c) convective cooling.

(a) *Cyclonic Cooling*

The cyclonic cooling may be divided into two types as frontal and non-frontal. In frontal type of cyclonic cooling, the circulation forces the air up over a frontal surface. The lifting and cooling of air is gradual thus producing moderate rainfall rates often of quite long durations. The non-frontal type of cyclonic cooling results from the convergence and subsequent lifting of air which accompanies a low-pressure area. Non-frontal cyclonic precipitation of extratropical origin produce rains (or snows) of moderate intensity but of fairly long durations. This type of storm may last from 24 to 72 hrs. and deliver a total of 50 to 150 mm of precipitation in that time. Non-frontal cyclonic precipitation of tropical origin may deliver as much as 375 mm of rain in a period of 12 to 24 hrs.

(b) Orographic Cooling

Orographic cooling occurs when the air is lifted up a slope to low pressures corresponding to higher elevations. Clouds first form and then precipitate if the dew point is reached. Of particular interest in the case of orographic cooling is what is known as the *rain shadow* or deficiency in rainfall which may occur on the leeside of the mountain slope because or removal of moisture by high mountains. Such a rain shadow region is found to the east of western ghats in india. In general, the windward side of mountains are more cloudy and rainy and have smaller temperature variations. On the leeward side, however, the climate is drier, sunnier, and more variable in temperature.

(c) Convective Cooling

When vertical instability of moist air is produced by surface heating, a convective current results. Convective precipitation is of very short duration, rarely exceeding one hour, covering limited areas of 20 to 50 km². But the intensities are so high that the total precipitation may amount to 75 to 100 mm in a duration of ½ to 1 hour.

Condensation

Under certain controlled conditions in the laboratory, lowering of the temperature of air does not produce condensationl investigations have shown that air, from which all foreign particles have been removed, can be cooled until the relative humidity is as much as 1000 percent before the droplets of liquid water form. However, condensation takes place even when the air is not saturated provided the air contains small particles of substance that have an affinity to water. These hygroscopic particles are called *condensation nuclei*. The size of these nuclei is of the order of 0.0001 to 10 microns. The condensation is hastened if the clouds possess

condensation nuclei in large quantities. Otherwise, though sufficient moisture is present in the clouds, they may not precipitate.

Growth of Droplets

The saturation vapour pressure over a curved surface is more than
over a horizontal surface. As the condensation nuclei attract water,
the size of the droplet increases making it less curved, and further
condensation therefore requires less super saturation. On the other
hand, the water attracted by the nuclei goes into solution and the
attraction for water decrease. These two effects of hydroscopy and
curvature, which counteract with each other in the initial stages of
the growth of droplet become negligible after the droplet has
attained a radius of 1 micron. Computations have shown that under
normal conditions, it takes a few seconds for the nucleus to grow to
a droplet of 10 microns, a few seconds for the nucleus to grow to a
droplet of 10 microns, a few minutes to reach 100 microns, 3 hrs to
grow to 1000 microns (1 mm) and about 24 hr to become a small
raindrop (of diameter 3 mm). It follows then that pure condensation
is incapable of producing rain drops though it may well produce
oversize cloud droplets. As the droplets gain mass they begin to have
a motion relative to their host cloud. This relative motion has the
effect of transporting additional water vapour into the neighbour-
hood of the droplet. It is known as the *ventilation effect*. When the
water droplet moves relative to the cloud, it may collide with other
droplets and grow further in size. This phenomenon is called
coalescence. As a drop grows to a diameter of 7 mm the fall velocity
increases to about 10 m/s. At such high speeds the drop flattens and
breaks up into drops of the size of small rain or drizzle. Each of these
will fall faster than the cloud droplets, with the result the coalescence
processes are repeated. It has been shown theoretically that droplets
of radius less than 18 microns cannot grow by coalescence, droplets
of radius less than 60 microns of nearly uniform size are relatively
stable against coalescence and droplets with radius greater than
80 microns of nerrly shows that the relatively a few large drops in a
cloud are most important in the formation of precipitation. This has
been the basis for "cloud seeding" or "artificial rain making". Dry
ice and silver iodide have been the common materials used as seeds
to accelerate the coalescence process. There is a varying degree of
success reported about artificial rain making, but the technique is
yet to be accepted for wider use.

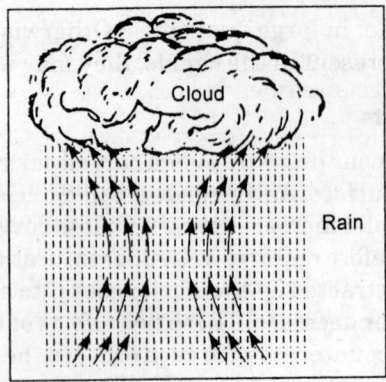

Fig. 26.1: *Convectional rainfall.*

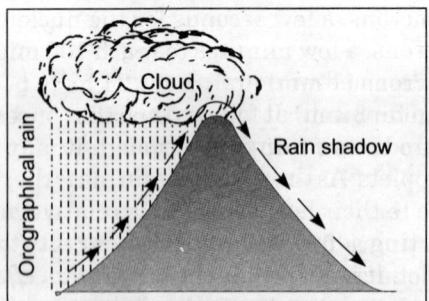

Fig. 26.2: *Orographic rainfall.*

TYPES OF PRECIPITATION

Precipitation is often classified according to the factors responsible for lifting and subsequent cooling. Broadly speaking there are three types of precipitation. They are:

(i)   Cyclonic precipitation.
(ii) Convective precipitation.
(iii) Orographic precipitation.

The mechanisms of cooling have already been discussed in the previous section. In nature the effects of these types of cooling are often inter-related and it may not always be possible to identify any one particular type of precipitation, as shown in Figs 26.1 and 26.2.

FORMS OF PRECIPITATION

The number of different forms of precipitation is very large. Only common types are described below.

Drizzle

It is a fine sprinkle of very small and rather uniform water drops with diameter between 0.1 and 0.5 mm. The drops are so small that they seem to float in the air. To qualify as *drizzle*, also sometimes called the *mist,* the drops must not only be small, but they must also be very numerous. The intensity of drizzle exceeds rarely 1 mm/hr.

Rain

Rain is precipitation of liquid water, in which the drops are generally larger than 0.5 mm in size.

Glaze

The ice coating formed when rain or drizzle freezes as it comes in contact with cold objects at the ground is called as *glaze*. It can occur only when the air temperatue is below 0° C.

Sleet

When rain drops are frozen while falling through air below 0° C, transparent globular grains of ice known as *sleet* or *ice pellets* are formed. The pellets are generally between 1 mm and 4 mm in diameter.

Snow

Precipitation in the form of ice cystals is called *snow*.

Hail

Balls or irregular lumps of ice over 5 mm in diameter is called *hail*. Hail stones are generally composed of alternating ice and opeque snow like layers as a result of repeated ascents and descents within the cloud during their formation. Hail occures almost exclusively in voilent or prolonged thunderstorms.

Dew

Dew forms directly by condensation on the ground, mainly during the night when the surface has been cooled by outgoing radiation.

Measurement of Rainfall

27

Rainfall is generally measured at a place in mm of water with the help of standard rain-gauges. The quantity of rainfall at any place during a given period is generally calculated on the basis of the depth of water which would accumulate on a level surface. It is interested to mention here that if there is no loss of water in any way such as percolation, evaporation, etc. then full rain water can be used which will be economical.

The rainfall is recorded by the weather observation stations, located in every part of the country with the help of rain gauges. These water observation stations record daily, monthly and annual rainfall of the area in which they are located. These data or records should be taken into account while computing the run-off calculations specially for a particular area.

Symon's rain gauge: Sometimes the rainfall is measured by Symon's rain-gauge due to its accuracy. This rain-gauge mainly consists of a graduated glass cylinder fixed with a funnel of the same diameter at the top. This rain-gauge is fixed on a masonary block in an open space, where the rain-water which enters the gauge is not affected in any way. This gauge gives the rainfall directly up to accuracy of 10 mm as shown in Fig. 27.1.

Standard rain-gauge: This rain-gauge consists of a cylinder 20 cm in diameter and its height is 60 cm. Its upper end is fitted with a funnel which discharges the rainfall water into a receiving tube, whose cross-section is 01/10th that of the cylinder. The height of water in the receiving tube is measured by means of measuring stick. The reading of the stick divided by 10 is the rainfall. This gauge gives reading up to 0.1 mm accuracy.

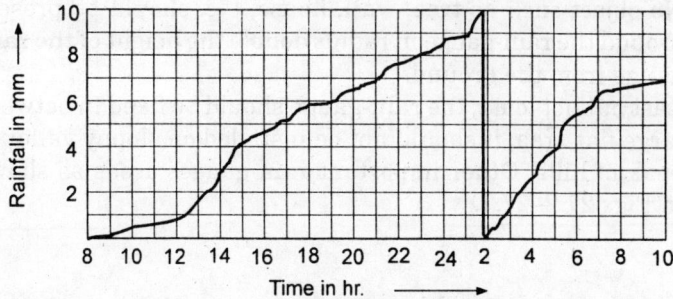

Fig. 27.1: *Chart from siphon raingauge with siphoning.*

Nowadays new standard type of nonrecording rain-gauge is mostly used in India. It mainly consists of a collector, with a gun metal rim, a base, and a polythene bottle. Both the collector and the base are made of fibre glass reinforced polyester. Its collector has a deep set funnel. The following table shows the combination of rain-gauge parts.

Collected area cm²	Base size	Bottle capacity in litres	Normal rain-gauge capacity in mm
200	Small	2	100
200	Small	4	200
100	Small	2	200
100	Small	4	400
200	Large	10	500
100	Large	10	1000

Now we come to the point is that the standard rain-gauge has a slight taper with a narrower portion at the top. The apertures of the collectors are 100 or 200 cm² in diameter which are interchangeable bases are also used. The smaller base is used with all sizes of rain-gauges except the largest. As we are well-known that there are three sizes of polythene bottles, viz. 2, 4 and 10 litres of water are used.

Important points are generally considered while selecting the site for fixing the rain-gauge.

1. The side should be properly protected from strong winds, as these may cause the eddies and disturb the rainfall in the rain-gauge, which may result in wrong collection of rain water.

2. No object such as tree, wall, house, etc. should be present around the rain-gauge in radius double the height of the rain-gauge from the ground.
3. Last but not least, the rain-gauge should be fixed in between large flat area, it should not be installed on sloppy valley or peak of hills. Other important rain-gauges areas as shown in Figs 27.2–27.7.

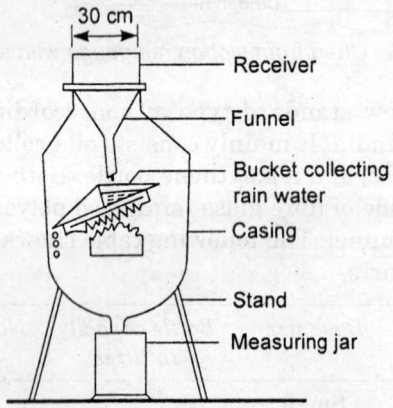

Fig. 27.2: *Tipping bucket type recording rain-gauge.*

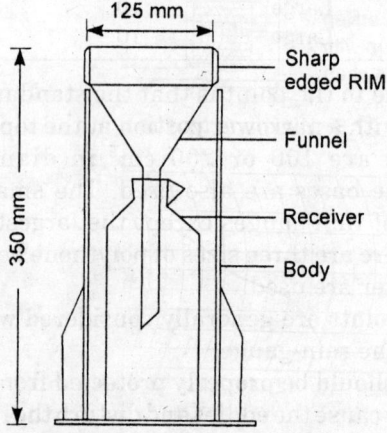

Fig. 27.3: *Simons rain-gauge.*

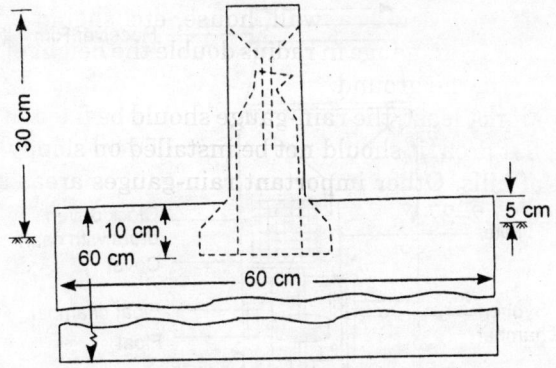

Fig. 27.4: *Installation of rain-gauge.*

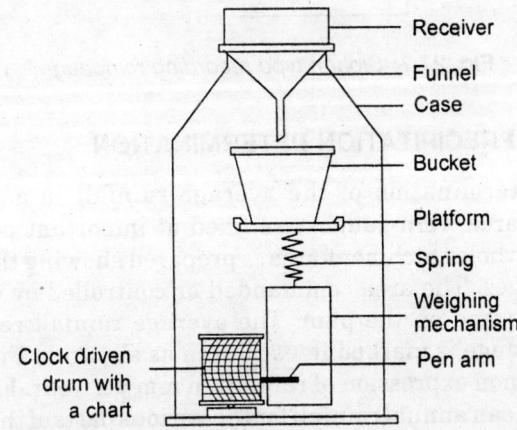

Fig. 27.5: *Weighing type recording rain-gauge.*

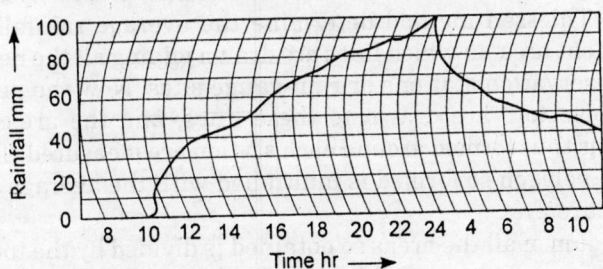

Fig. 27.6: *Chart from recording raingauge with reverse mechanism.*

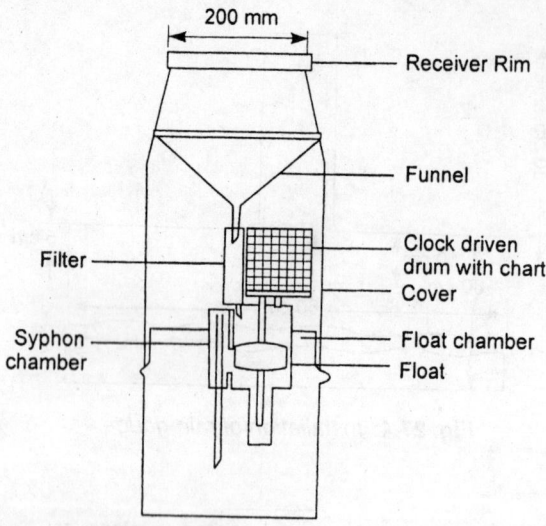

Fig. 27.7: *Siphon type recording raingauge.*

AVERAGE PRECIPITATION DETERMINATION

For the determination of the average rainfall in a particular catchment area, rain-gauges are fixed at important points. The plan of the whole catchment area is prepared showing the position of rain-gauges. The area commanded or controlled by each rain-gauge is marked on the plan. The average rainfall recorded by each rain-gauge is marked in each area as shown in Fig. 27.8.

The common expression of rainfall in mm per year. Figure 27.9 shows the mean annual rainfall for the various parts of the country. All the lines of equal rainfall are known as *isohyet* and the map in which these lines are shown in Fig. 27.9 is known as *isohyetal map*.

The Thiessen method determine the average rainfall over a catchment area, in which lines are drawn joining all the neighbouring observation stations or rain-gauge sites. Now the perpendicular bisectors are drawn to these lines, and the areas of the polygons thus formed around each station are measured. The area of each polygon so formed is multiplied with the average rainfall of that area.

The sum of all the areas so obtained is divided by the total area to get the average rainfall in the whole catchment area.

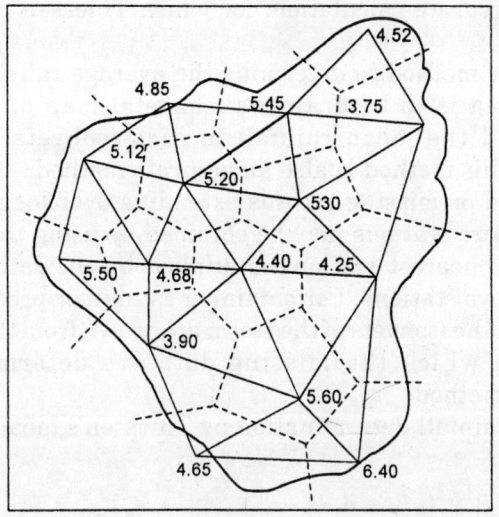

Fig. 27.8: *Thiessen method of determining average rainfall in a particular catchment area.*

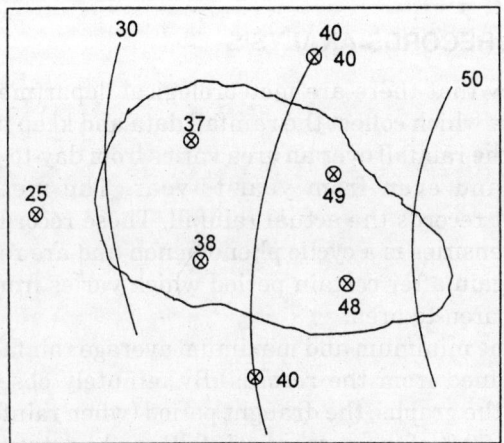

Fig. 27.9: *Isohyetal map.*

At sometimes, the average of all the rainfall recorded by various rain-gauges is determined, to get the average rainfall in the whole catchment area, but in fact, this is a rough method and is not suitable for accurate calculation, for which Thiessen method is used.

The another method to determine the average rainfall in the catchment area is to be drawn is isohyetal map of the area concerned and the mean rainfall in each isohyets area are determined. This method is also an accurate method.

Unidentified or missing records extending over long periods, from one or more stations may be obtained by using the annual figures for the nearest stations multiplied by the ratios of the means of the two stations. For obtaining rainfall or precipitation during a storm the isohyets of the storm are drawn from the known figures, from which the missing data are determined by interpolation method.

The mean rainfall determination by Thiessen's mean method is as follows:

$$P = \frac{A_1 P_1 + A_2 P_2 + A_3 P_3 + \dots\dots\dots + A_n P_n}{A}$$

where, P = Mean rainfall on the basin

A = Basin area.

RAINFALL RECORDS ANALYSIS

As we know that there are meteorological department's centres in every city which collect the rainfall data and keep its up to date record. As the rainfall over an area varies from day-to-day, month-to-month and even from year-to-year. The meteorological department records the actual rainfall. These records show that rainfall intensities is a cyclic phenomenon and are repeated over and over again after certain period which varies from region-to-region and area-to-area.

In fact, the minimum and maximum average rainfall quantities are datermined from the records. By minutely observing these values and the graphs, the draught period (when rainfall intensity is below 60–70 % of the average rainfall) can be determined. Extra water storage shall be required to meet the drought period.

Hydrogeologists or water-works engineers mainly require the rainfall records the purposes as follows:

(a) The duration of the dry weather and the availability of the water in that period will help in deciding the actual size of the

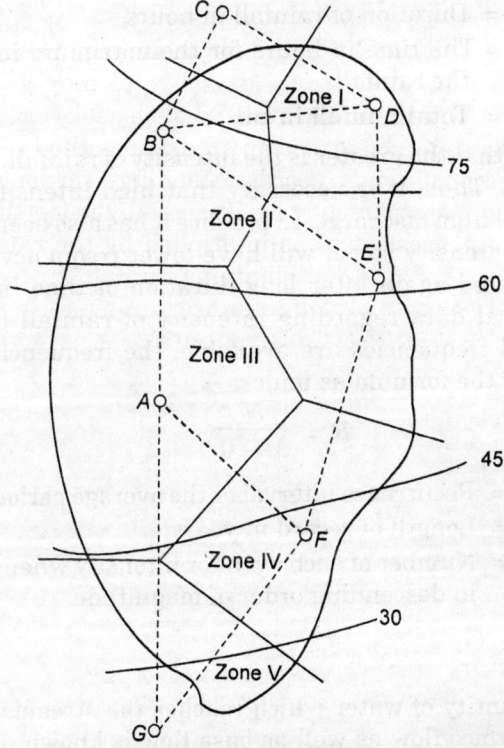

Fig. 27.10: *Depth-area-duration analysis.*

storage reservoir. For this also at least dry weather records of three successive dry years should be taken into account.

(b) The maximum intensity of rainfall will help in determining the maximum quantity of water, which will flow in the rivers, lakes, etc. The design of collection works, bridges, culverts and other structures will directly depends on the maximum quantity of run-off.

Relationship of Rainfall Intensity Duration Frequency

The rate of rainfall over a particular area and its duration are very important for several reasons. The intensity of rainfall in cm per hour is determined by the following formula.

$$i = \frac{T+1}{t+1} \times \frac{F}{T}$$

where, i = Intensity of rainfall in centimetres.

T = Duration of rainfall in hours.

t = The time in hours for the maximum intensity of the rainfall.

F = Total rainfall in cm.

It is a fact that the greater is the intensity of rainfall, shorter is the duration. Thus, it is necessary that high intensity rainfall will give maximum discharge. In practice it has also been observed that a high intensity storm will have lower frequency that is it will be repeated again after long duration of time interval. It sufficient local data regarding intensity of rainfall of various duration and frequencies are available, the frequencies can be calculated by the formula as under:

$$R = \frac{N}{M - 0.5}$$

where, R = Recurrence interval or the average period in years.

N = Length of record in years.

M = Number of each event or intensity when arranged in descending order of magnitude.

RUN-OFF

The total quantity of water which reaches the streams or rivers both from surface flow as well as base flow is known as *run-off*. The run-off mainly depends on the following factors:

(a) Area of the catchment.

(b) The degree of porosity of the soil of the catchment area.

(c) Slope and shape of the catchment area.

(d) Initial state of the catchment area with respect to the wetness.

(e) Duration of rainfall.

(f) Intensity of rainfall in catchment area.

(g) Obstruction in the flow of water due to trees, fields, gardens, bunds, etc. (Fig. 27.11).

EVAPORATION LOSS DETERMINATION

There are various empirical formulae for the determination of evaporation loss but generally used methods are following.

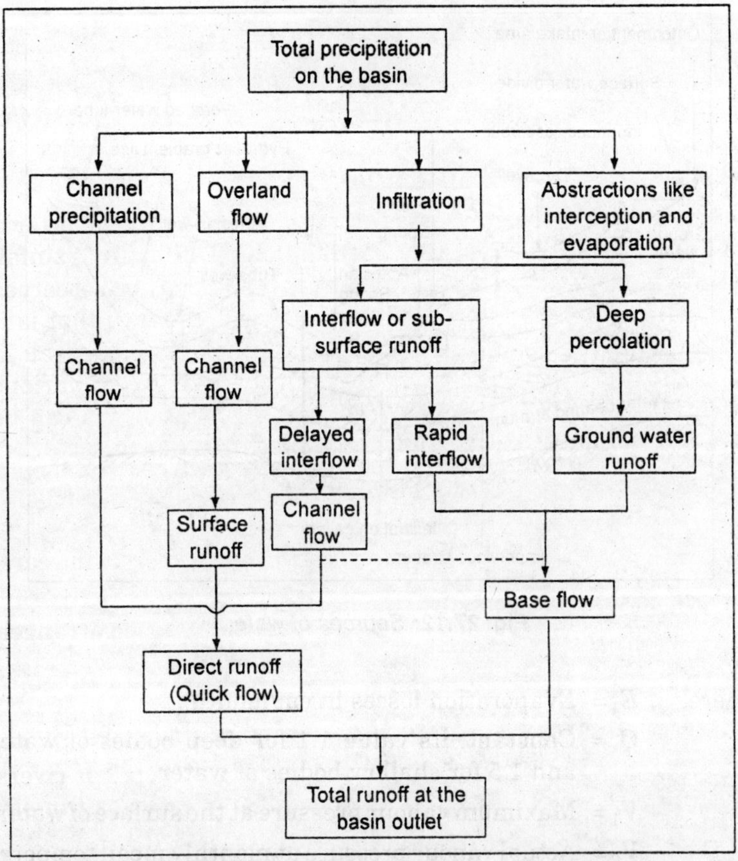

Fig. 27.11: *Schematic respresentation of run-off process.*

Rohwer's Formula

$$E_1 = C\,(1.465 - 0.00732P)\,(0.44 + 0.0732V\,)\,(V_1 - V_2)$$

where, $C = 0.75$, determined from the available evaporation
data.

$P =$ Atmospheric pressure in cm at 0° C temperature.
E_1, V, V_1 and V_2 have the same meaning as in the
following Meyer's formula.

Meyer's Formula

$$E_1 = C\,(V_1 - V_2)\,(1 + V/K)$$

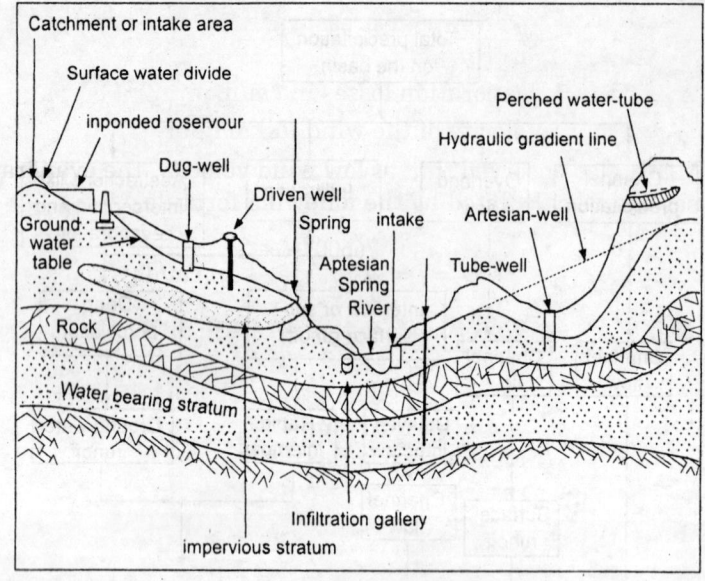

Fig. 27.12: *Sources of water.*

where, E_1 = Evaporation losses in cm/month.

C = Constant, its value 1.1 for deep bodies of water and 1.5 for shallow bodies of water.

V_1 = Maximum vapour pressure at the surface of water.

V_2 = Actual vapour pressure at monthly mean temperature and atmospheric relative humidity.

V = Velocity of the wind in km/hour.

K = Some constant whose value is up to 16.

MERI

Maharashtra Engineering Research Institute (MERI) at Bhatghar and Khadakwasla, have given the following formula for the evaporation losses calculation

$$E_1 = 0.4 + 0.04.V + 0.50\ (e_s\text{-}e_d)\ \text{cm/day}.$$

where, e_s = Mean vapour pressure in cm of saturated air at water surface temperature.

e_d = Mean vapour pressure in cm at dew point temperature.

E_1 = Evaporation losses in cm/month.

v = Velocity of the wind in km/hour.

MERI also observed that at low wind velocity, the evaporation losses can be calculated by the following formula.

$$E_1 = i = \frac{76600}{T^{3.369}} \, (e_s - e_d) \text{ cm/day.}$$

where, T = Water surface temperature at °C.

E_i = Evaporation losses in cm/month.

e_s = Mean vapour pressure in cm of sturated air at water surface temperature.

e_d = Mean vapour pressure in cm at dew point temperature.

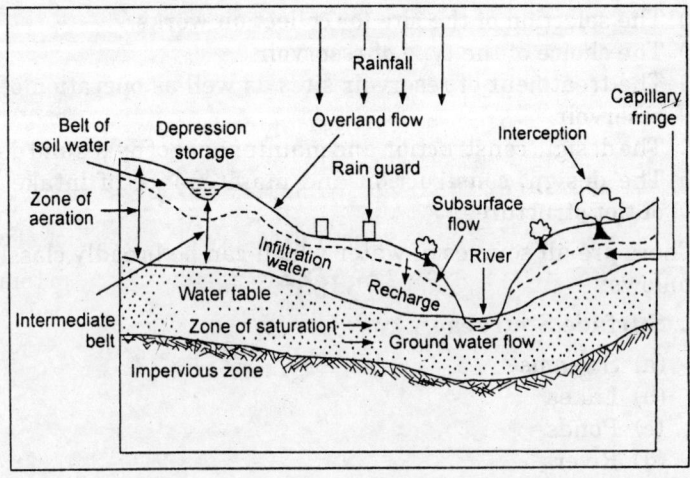

Fig. 27.13: *Generalised cross-section showing run-off terms.*

28 Water of Earth's Surface

There are various sources of water but surface sources of water are generally include rivers, streams, lakes, ponds, etc. Water yield of the surface sources vary from season to season. Important, reliability and quantity of water mainly depends on the following.

(a) Preparation and control of the catchment area.
(b) The selection of the site for collection works.
(c) The choice of the type of reservoir.
(d) The treatment of reservoir sites as well as operation of the reservoir.
(e) The design, construction and maintenance of dykes and dams.
(f) The design, construction and maintenance of intake and outlet structures.

There are all sources of water which can be broadly classified as follows:

1. Surface Sources

(a) Streams
(b) Lakes
(c) Ponds
(d) Rivers
(e) Imponded reservoir
(f) Stored rainwater and cistern

2. Ground Sources

(a) Springs
(b) Infiltration galleries
(c) Porous pipe galleries
(d) Wells

SURFACE SOURCES OF WATER

Streams

Streams are formed by the run-off in mountainous region. The rainy season provide much discharge than any season. Stream which dry up in summer and contain water only during rainy season are known as *rainy streams*. The water quality in streams is normally good except the water of first run-off. But sometimes run-off water is mixed with clay, sand and mineral impurities and these impurities can be removed in settling tanks up to certain extent. But as we know that the dissolved impurities require special treatments. The streams are the main source of water supply to villages of hills which are around them.

Lakes

Natural basins are sometimes formed with impervious beds in mountainous area. Water from springs and streams generally flows towards where basins and *lakes* are formed. In lakes, the quantity of water depends mainly on its basin capacity, catchment area, annual rainfall, porosity of the ground. Large lakes water quatity is good than that of small lakes. But lakes which are situated at high altitudes contain almost pure water which can be used without any treatment. The cities and towns which are near the lakes have easy availability of lake water, such as Nainital in Uttranchal Pradesh.

Ponds

Ponds are depressions in plains like lakes of mountains, in which water is collected during rainy season. Sometimes ponds are formed when much excavation is done for constructing kachha houses in villages, embankment for road and railways, and manufacture for bricks. Generally, the quantity of water in ponds is very small and contains large amount of impurities. In most backward villages people also take bath in dirty water of ponds. The water of pond cannot be used for water supply purposes due to its limited quantity and large amount of impurities.

Rivers

The motherland of rivers is hills. In mountains the quantity of water in rivers remain small, therefore at such places these are generally known by small rivers. When the rivers move forward, more and more streams combine in it and it increases its discharges. It results in growing rivers bigger and bigger as they move forward due to increase in their catchment area. Rivers are the only surface sources of water which have maximum quantity of water which can easily be taken, therefore, at a very ancient times the town and cities started developing along the banks of rivers. Mostly all the cities near rivers discharge their wastes into rivers and pollute the water of rivers, etc.

GROUND SOURCES

Sometimes ground water reappears at the ground surface in the form of springs. It can supply water for small hill towns. It is interesting to note that due to presence of sulphur in certain springs, they discharge hot water. Such hot water springs are only useful for taking bath for the cure of certain skin disease patients. The most common example of hot spring is at Rajgir of Bihar in India.

(i) When the surface of the earth drops sharply below the normal ground water table is generally called by *shallow springs* as shown in Fig. 28.1 (a).

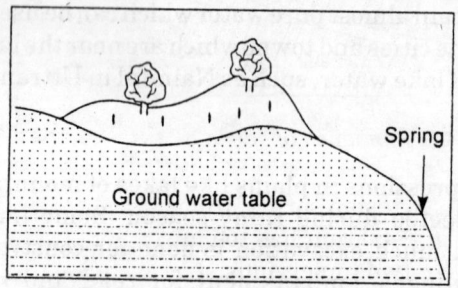

Fig. 28.1 (a): *Gravity or shallow springs.*

(ii) When due to an obstruction ground water is collected in the form of a reservoir and forces the water to overflow at the earth's surface is called by the name of *surface spring* as shown in Fig. 28.1 (b).

(iii) When the fissure in an impervious stratum allows artesian water to flow in the form of springs is known by the name of springs formed due to fault in the rock as shown in Fig. 28.2.

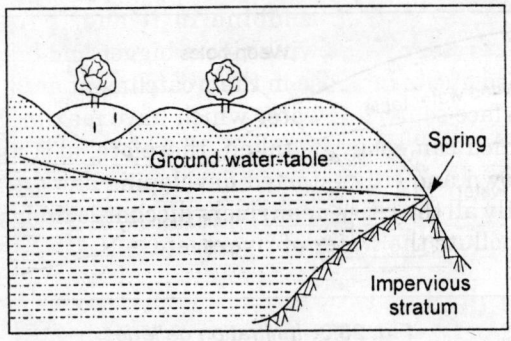

Fig. 28.1 (b): *Common type of spring (surface spring).*

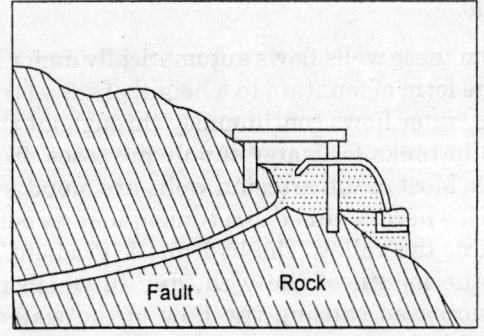

Fig. 28.2: *Formation of spring due to fault in the rock.*

Formation of "Artesian spring" from a deep seated spring. When the water-bearing stratum has too much hydraulic gradient and is closed between two impervious stratums, such type of springs are formed.

Water travelling towards lakes, rivers or streams can be intercepted by digging a trench or by constructing a tunnel with holes on sides at right angle to the direction of flow of underground water. These underground tunnel used for tapping underground water near rivers, lakes or streams are called *infiltration galleries* as shown in Fig. 28.3.

These galleries yield as much as 1.5×10^4 litres/day/metre length of the infiltration gallery. Infiltration galleries are also known by the name of *horizontal walls*.

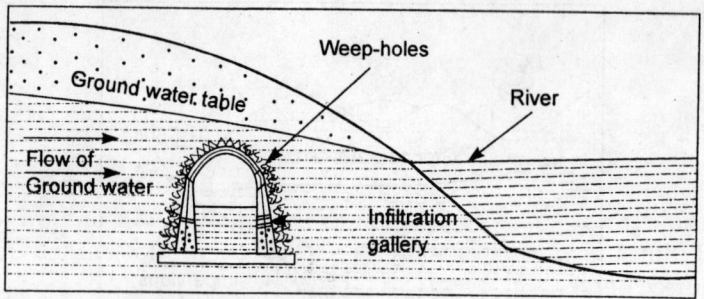

Fig. 28.3: *Infiltration galleries.*

ARTESIAN WELL

Artesian Wells

The water from these wells flows automatically under hydrostatic pressure in the form of fountain to a height of 2.5 metres. In some artesian wells water flows continuously throughout the year and can be stored in tanks for water supply purposes, from 4,500 to 1,14,000 litres. Most of the artesian wells are found in the valley portion of hills, where confined stratum aquifer on both sides are inclined towards the valley. The hydraualic gradient line passes much above the mouth of the well, due to which hydrostatic pressure is increased causing the flow of the water under its pressure (Fig. 28.4).

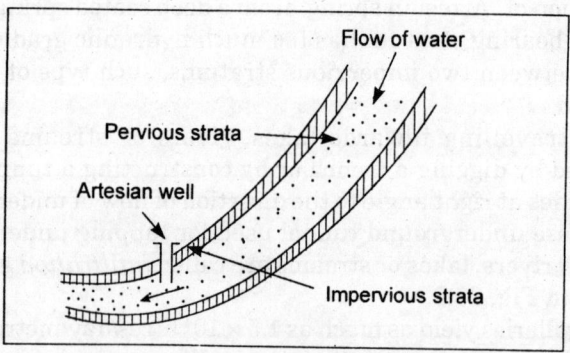

F₁. 28.4: *Artesian well.*

Dug Wells

Dug wells are shallow wells. They are up to 20 metres deep and 1 to 4 metres in diameter, and are suitable for small discharge of about 20 cu m/hour. To construct these wells, first of all a well is constructed at the site over which of up to a height of 1.5 metres, masonary work is done. The earth now is excavated from the inside of the cure by means of pick-axes and showel. Masonary work will continue, as excavation proceeds till the well is sinked up to required depth.

As it is cheap to construct dug wells, so these are very popular in rural areas as well as in small towns. All types of small demands can be met by these wells because these are large in diameter and act as small reservoirs. These are also easy to disinfect from any kind of contamination and it is must be done due to its nature to be of poor sanitary conditions.

The quantities of water which can be withdrawn from these wells are limited due to their limited capacity to store water and also depends on the critical velocity for the soil. If withdrawal of water from these wells is done, the soil particles will be disturbed and will lead to sinking of the well steining.

If it is needed to increase the yield from these wells, we should provide a 8–10 cm diameter bore hole in the centre of the well, which will trap more quantity of water from the aquifer below the hard stratum. These are well described under the heading of open wells (Fig. 28.5).

Open Wells

As it is mentioned above, the open wells are constructed up to a limited depth and their yield is also limited. The yields of open wells are mainly increased by providing a 8 to 10 cm diameter bore hole in the centre of the well (Fig. 28.7 (a) and (b)).

If the open-well is resting on soft stratum, and the mota layer (mota layer = the layer of clay, kankar, cemented layer and other hard layer existing a few metres below the first water table in the sub-soil) is not at greater distance, so we must do to sink the well further.

Geological formation of the ground sub-soil controls and define the difference between shallow well and deep well. The shallow well draws its water only from the topmost water-bearing stratum,

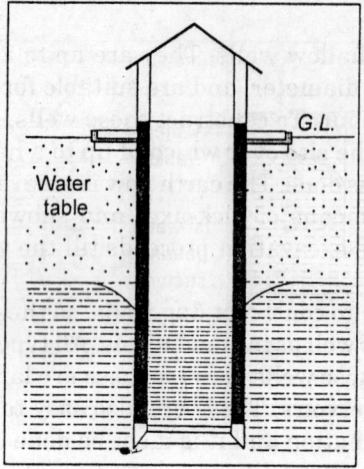

Fig. 28.5: *Dug-well.*

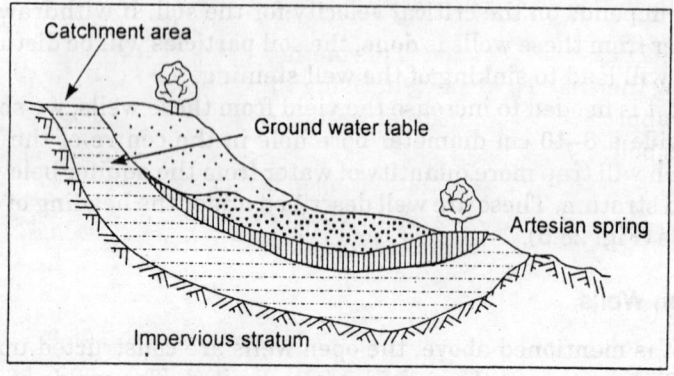

Fig. 28.6: *Deep-seated spring.*

whereas deep well draws its water from more than one water-bearing strata. It is important to mention that shallow wells may be deeper than the deep wells.

As we know that the yield of open well can be mainly measured by three methods.

1. Theoretical method.
2. Actual pumping method.
3. Recuperating test method.

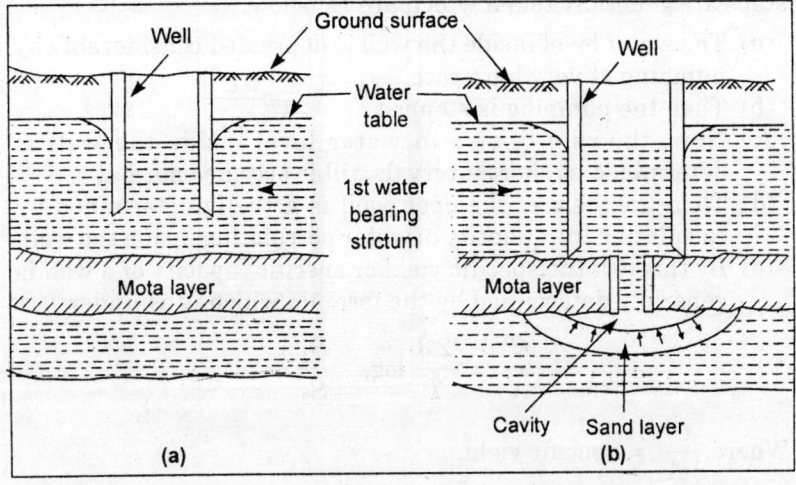

Fig. 28.7 (a): *Existing open well.*   **Fig. 28.7 (b):** *Improved open well.*

1. *Theoretical-method*: This method is an important method to know the yield by open well. Using this method, we can calculate the approximate quantity of water percolating in the well which is as below:

If, A = Area of aquifer opening into well.
       V = Velocity of the water percolating in the well.
       B = Permeability constant.
       Q = Yield of the well.
Q (Yield of the well) = AVB

2. *Actual pumping method*: In actual pumping method the following steps are generally taken:
  (a) The water is withdrawn from the well at very high rate.
  (b) The rate of pumping is reduced and adjusted by trial and error till the water level in the well remains constant.
  (c) At this stage, the rate of percolation inside the well and the rate of pumping become equal.
  (d) This results the determination of rate of pumping which is the yield of the well.
  (e) This test is generally done during summer season to determine the dry weather yield of open well.

3. *Recuperating test method*: The yield of open well is also determined by determining the rate of percolation inside the well.

This method is also highly applicable in this method. The following steps are generally taken which are as below:

(a) The water level inside the well is depressed considerably by pumping at very high rate.

(b) Then the pumping is stopped.

(c) Now, the rate of rise in water level inside the well is determined at short intervals, till it becomes normal.

(d) Then the yield of the open well is actually determined by calculating the quantity of water percolating inside the well.

(e) By this test the specific yield or specific capacity of a well is generally determined by the formula which is as under:

$$\frac{C^1}{A} = \frac{2.3}{T} \log_{10} \frac{S_1}{S_2}$$

Where, $\dfrac{C^1}{A}$ = Specific yield.

S_1 = Depression head in the well at the time just after the pumping was stopped.

S_2 = Depression head in the well at the time T, after the pumping was stopped.

T = Time after which the measurement S_2 was taken.

By knowing the value of $\dfrac{C^1}{A}$ = the discharge Q of the well can be determined by the formula which is as under:

If, $\quad\quad A$ = Cross-sectional area of the open well.

Q = Yield of the well.

S = Depression head.

Then, $\quad\quad Q = \left(\dfrac{C^1}{A}\right) = AS$

We know that the maximum discharge from the ordinary open wells is between 4 to 5 litres/second. It is very low yield so open wells are useful for small locality or private estates.

Thus, it is not economical to install pumps in open wells, due to their low yield.

Due to rise in population, greater demand for supplying treated water for the use of domestic purpose, agricultural purpose and drinking purpose, so, more yield of water is needed. To meet this requirement, we construct tube wells of different types which are mainly as under:

1. Strainer type tube wells.
2. Cavity type tube wells.

As tube wells having depth from 50 to 500 m from the surface of the earth. The yield of the average tube well is about 50 litres/second, and the maximum yield may be about 200 litres/second.

Strainer Type Tube Wells

Most of the State Government Departments construct only this type of tube wells so it is in maximum use.

In this type of tube well a fine screen or strainer is placed against water-bearing stratum. Figure 28.8 as shown, the diameter of the outer shell of the tube well varies from 15 to 100 cm and the diameter of the pipe through which water draw varies from 2.5 to 90 cm. Gravel is sometimes packed outside the strainer pipe, approximately 8 cm thickness, which helps in checking the entrance of sand particles in the tube wells and increase the yield of tube well. This type of tube well is not suitable for very fine sandy water-bearing formation. Strainer type tube well can be constructed by taking following steps:

1. For the construction of such tube wells, we do boring and the diameter of casing is between 5 to 10 cm larger than the diameter of the well pipe is lowered.
2. While doing boring work and lowering the casing pipe.
3. The samples of the soils obtained at different are mainly noted down.
4. Actual records of the water-bearing formation are also prepared with respect to their length and depth.
5. When boring is done, the details of the soils at different depths are prepared.
6. Now we cut the strainer and blind pipes to the actual lengths for various stratum depending mainly on their thickness.
7. Well pipe assembly is done, as according to the stratum depth chart and the well pipe is lowered in the casing pipe such that blind pipe rest on dry stratum and strainer pipes against water-bearing formation.
8. When there is lowering of the tube well is completed, after this casing pipes are withdrawn.

9. To avoid the sand filling from the bottom side or prevent the well sinking, the bottom of the tube well is generally cemented (Fig. 28.8).

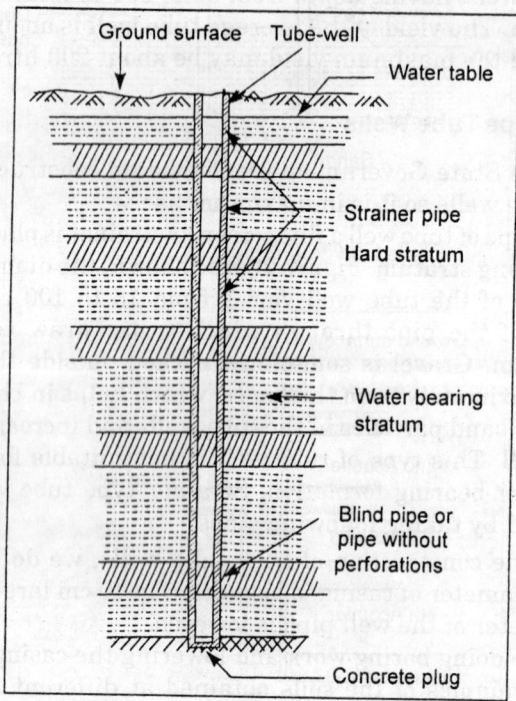

Fig. 28.8: *Section through a strainer type tube well.*

Cavity Type Tube Wells

This type of tube well are not generally commonly used. There is not suffient yield from this tube well and tube wells mainly draw their yield from their bottom aquifer only. In this type of tube well, the water may be drawn from the first, second, third aquifer.

1. In this type of tube well, blind pipes throughout its length is essential (Fig. 28.9).
2. Just after construction fine sand, silt come out along with the water which causes formation of a cavity in the bottom, as it is done in the beginning.

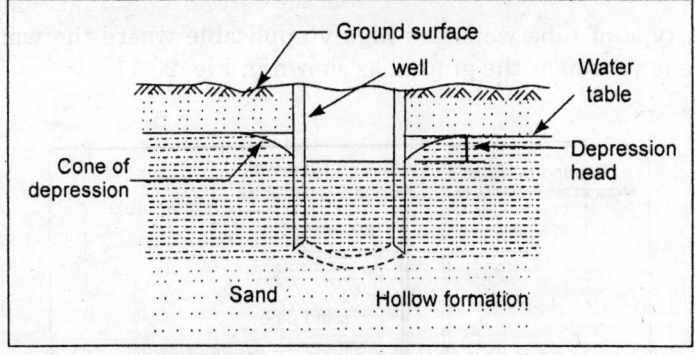

Fig. 28.9: *Cavity of hollow formation in well.*

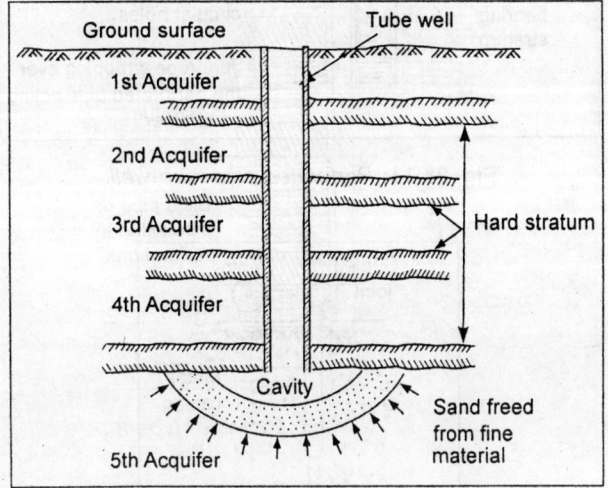

Fig. 28.10: *Cavity type tube-well.*

3. With the water velocity increase the spherical area of the cavity increases.

4. When the area of the cavity increases the critical velocity decreases which causes reduction in yield, this prevents the entry of the sand particles without coming water.

5. Thus, we do that, we increase the size of the cavity which naturally increase the yield of these wells.

Perforated Pipe Tube Wells

This type of tube wells are highly applicable where the water-table is very near the ground as shown in Fig. 28.11.

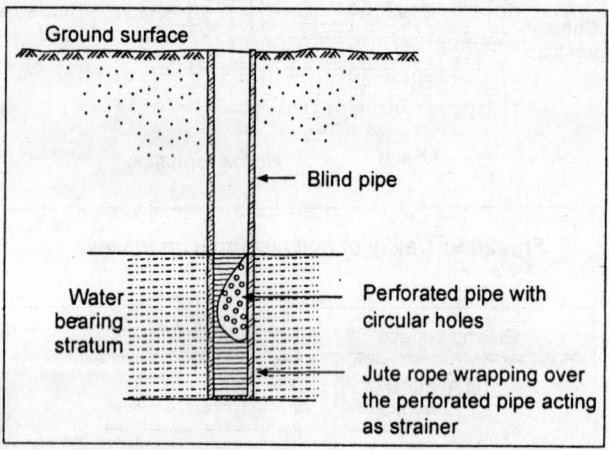

Fig. 28.11: *Perforated pipe tube-well.*

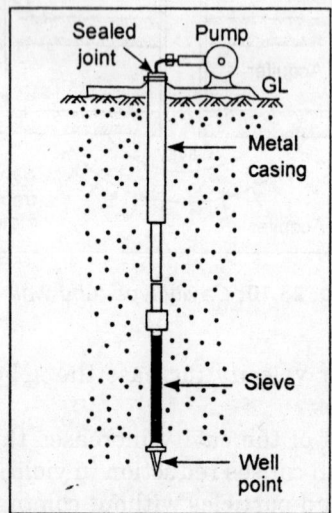

Fig. 28.12: *Driven well.*

1. In these wells the pipes are made perforated by drilling holes in them. These holes are covered by wrapping coconut jute rope over the pipe. This jute rope acts as strainer to prevent the flow of sand particles.
2. Perforated pipe tube wells are commonly used during the construction of canals, dams and other irrigation structures for obtaining the water during construction and also for doing the dewatering work and lowering the water-table.
3. It is important to note that in India these types of tube wells are not commonly used for water supply for various purposes.

Percussion Wells

This type of tube well is constructed by driving a casing pipe of 2.5 to 15 cm in diameter. The lower portion of casing pipe in water bearing formation is perforated (Fig. 28.13).

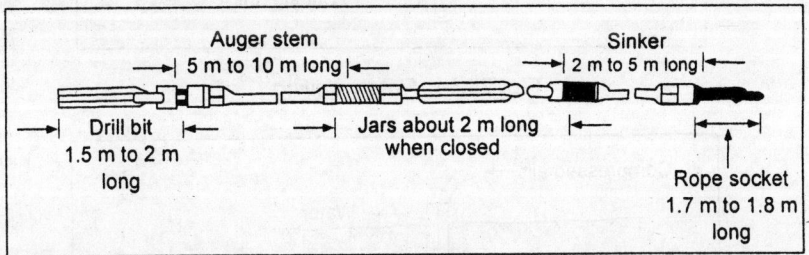

Fig. 28.13: *Percussion drilling.*

1. The bottom end of the casing pipe in pointed is commonly known by the name of well pointer drive point.
2. The perforated portion of pipe is covered with fine wire gauge to present the passage of soil and sand particles inside the well.
3. This type of tube well is most suitable for the domestic purpose only because the discharge of this tube well is very small.
4. For obtaining more quantity of water numbers of driven wells should be used.
5. The casing pipe is placed up to 12 metres deep is driven in the ground by means of hammer or by water jet.
6. After the depth of 12 metres, the water can be taken out by means of pumps.

Some Other Examples

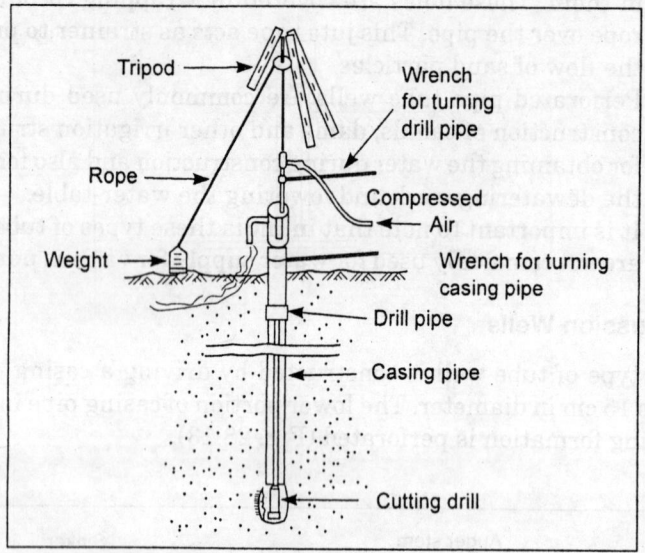

Fig. 28.14: *Rotary drilling.*

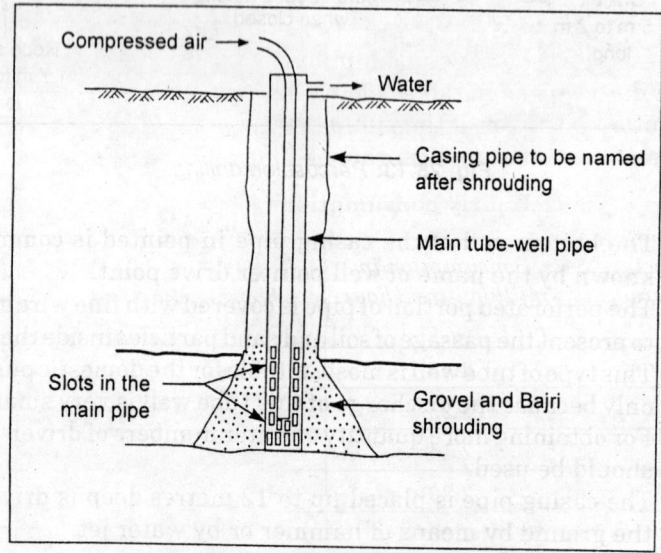

Fig. 28.15: *Slotted type tube well.*

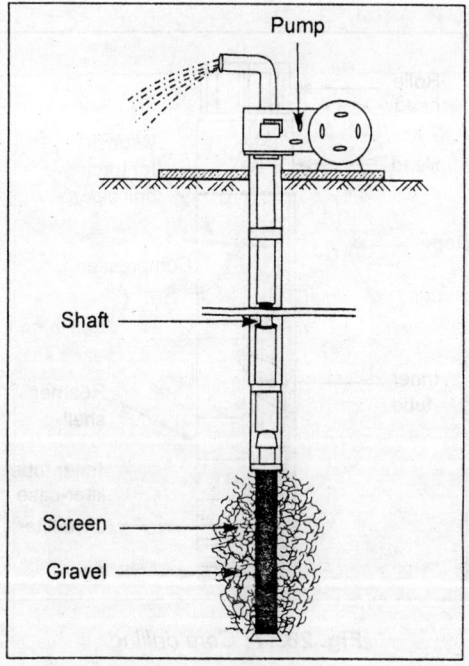

Fig. 28.16: *Development of wells by gravel packing.*

The estimated hourly consumption of water for a town for daily use per day as given in table below. Determine the capacity of the distribution reservoir, if the pump installed can supply the water in the reservoir at a uniform rate of 1.50 m³/second.

Hourly consumption of water

Time in hours	Consumption in million litres / hour	Time in hours hour	Consumption in million litres /
1	2.45	7	3.50
2	2.25	8	5.25
3	2.14	9	6.10
4	2.30	10	6.55
5	2.55	11	7.25
6	2.60	12	7.35

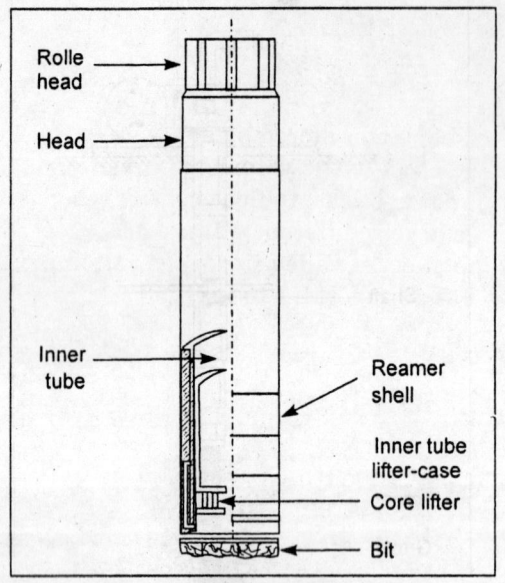

Fig. 28.17: *Core drilling.*

Hourly consumption of water *(Contd.)*

Time in hours	Consumption in million litres / hour	Time in hours hour	Consumption in million litres /
13	7.55	19	7.30
14	6.35	20	7.25
15	5.95	21	5.65
16	5.75	22	4.50
17	5.65	23	3.70
18	7.45	24	2.85

Solution

Rate of Pumping = 1.50 m³/sec.
                = 1.50 × 60 × 60 × 1000 litres/hour.

or,
$$= \frac{1.50 \times 36}{10} = \frac{27}{5} \text{ million litres/hour}$$

$$= 5.4 \text{ million litres/hour}$$

The volume of water stored by a stream within its banks may be small in relation to the volume of water during floods, when the stream is out of banks. In flood control reservoirs the effect of valley storage is very different water discharges. In flood control reservoirs the effect of valley storage is very important.

It is also not unimportant to say that as it is uneconomical to provide storage for the entire flood volume, as it is necessary to discharge some flow during filling period. When the rate of discharge has been fixed, the flow hydrograph, modified by discharge, determines the size of the reservoir.

29 | Sedimentation of Reservoir

All the materials deposited on the bed of the reservoir, then they all are transported by flowing water, this is known as *sedimentation*. The soil particles eroded by the water from the surface of the ground, are generally carried by the water to the reservoir where they settle in its bed.

The sedimentation problem is very serious problem in arid regions where the ground is not covered by vegetations results in excess erosion by heavy rainfall. The sedimentation percentage carried by the streams varies from stream to stream and characteristics of the catchment area. Due to gravitational force, the sediment particles try to settle the bottom of the river. Some particles remain in suspension position due to turbulent flow and these particles reach in the reservoir and settle down in the reservoirs bed and they are called *silting*.

The capacity of the reservoir reduces, as the quantity of silting increases. The most effective method to control sedimentation is by watershed control. To control the entry of the suspended particles in the reservoir, special check dams and debris barriers may be constructed at suitable places.

As there are various methods to reduce the quantity of silt, but some of them are as follows:

Type of soid	Voids
Sands and gravels of fairly uniform size and moderately compacted.	35 to 40%
Well graded and compacted sands and gravels.	24 to 30%
Sandstone	14 to 30%

(Contd.)

Type of soid	Voids
Chalk	14 to 30%
Schist and gneiss	0.02 to 2%
Slate and shale	0.5 to 8%
Limestone	0.5 to 17%
Clay	44 to 47%
Top soils	37 to 65%

Classification of Soils

Materials	Particle sizes in mm
Gravel	Larger than 2.0 mm
Very coarse sand	Between 1.0 to 2.0 mm
Coarse sand	Between 0.5 to 1.0 mm
Medium sand	Between 0.25 to 0.5 mm
Fine sand	Between 0.125 to 0.25 mm
Very fine sand	Between 0.07 to 0.125 mm
Cilt and clay	Finer than 0.07 mm

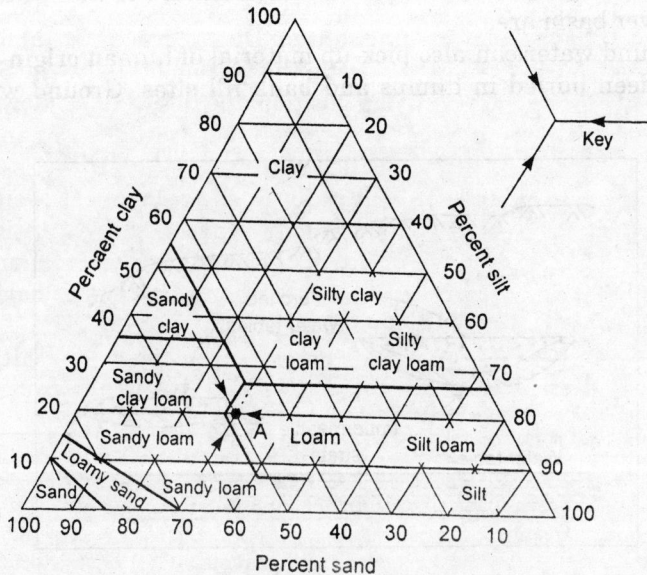

Fig. 29.1: *Triangle of soil textures for describing various combinations of sand, silt, and clay (after soil survey staff).*

 # Ground Water or Subsurface Water

Ground water is important to the ecosystem because it provides a reservoir for storing water slowly replenishing the lakes in the form of base flow in the tributaries. It is also a source of water supply for industries, domestic and agricultural sectors. Shallow ground water also provides moisture to plants. As water passes through subsurface areas, some substances are filtered out, but some materials in the soils get dissolved or suspended in the water, salts and minerals in the soil and bedrock are the source of what is referred to as "hard" water a common feature of well water in any river basin area.

Ground water can also pick-up material of human origin that have heen buried in dumps and bank fill sites. Ground water

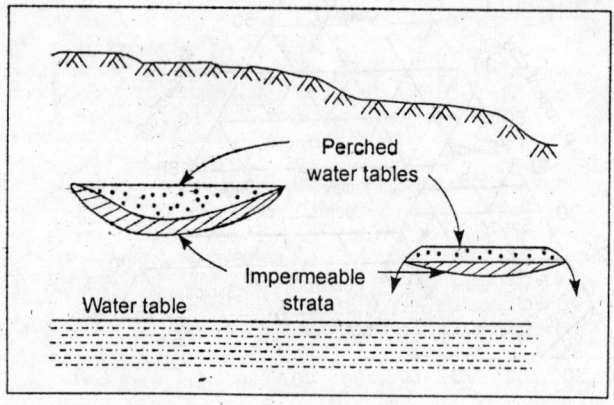

Fig. 30.1: *Perched water table.*

pollution can occur in both urban, industrial and agricultural areas. Conservation and inspection of ground water is essential to protect the quality of the entire water supply consumed by population world-wide because the underground movement of water is believed to be a major pathway for the transparent of pollution to the reservoirs. Ground water may discharge directly to the reservoirs or indirectly as base flow to the tributiries. As water slowly percolates through the ground it can pick-up dissolved materials have been burried or soaked into the ground. Figure 30.1 shows perched water aquifer and Fig. 30.2 shows effluent and influent streams.

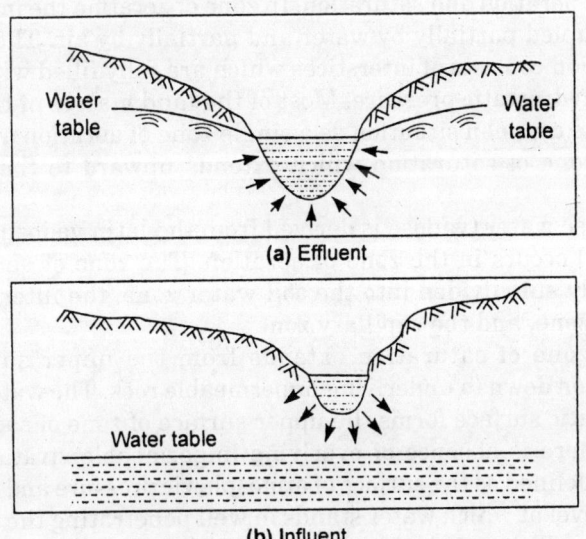

Fig. 30.2: *Effluent and influent streams.*

31 Vertical Distribution of Ground Water

The occurrence of ground water as subsurface may be divided into zones of aeration and saturation. In zone of aeration the interstices are occupied partially by water and partially by air. The zone of saturation consists of interstices which are fully filled with water under hydrostatic pressure. Most of the land masses of the earth have one common situation as a single zone of aeration overlies a single zone of saturation which extends upward to the ground surface.

Vadose water (vadose is derived from the latin vadosus means shallow) occurs in the zone of aeration. The zone of aeration is generally sub-divided into the soil water zone, the intermediate vadose zone, and the capillary zone.

The zone of saturation extends from the upper surface of saturation down to underlying impermeable rock. The water table, or phreatic surface forms the upper surface of zone of saturation when there is absence of overlying impermeable strata. Water table is defined as the surface of atmospheric pressure and appears as the level at which water stands in well penetrating the aquifer. Due to capillary attraction, saturation extends slightly above the water table, however, water is held here at less than atmospheric pressure. Water occurring in the zone of saturation is generally called by the name of ground water or phreatic water.

ZONE OF AERATION

Zone of soil and water: In general, water in the soil-water zone exists at less than saturation, but sometimes due to heavy rainfall or irrigation practice, the excess water reaches the ground surface, but it happens temporarily. This zone primarily extends from the ground surface down through the major root zone. It has different thicknesses which are affected by soil type and vegetation. The extensive studies made by soil scientists and agriculterists on soil moisture distribution and movement because of the agricultural importance of soil water in supplying moisture to roots.

The soil-water zones contention of the amount of water depends primarily on the recent exposure of the soil to moisture. In fact a water vapour equilibrium tends to become established between the ambient air and the surfaces of fine-grained soil particles. Hygroscopic water is thin films of moisture is resulted which remain adsorbed on the surfaces. Water also forms liquid rings surrounding contacts between grains as in coarse-grained materials and where additional moisture is available, which is shown in Figs 31.1 and 31.2.

This water is only held due to surface tension forces and is sometimes referred to as capillary water. Under the influence of gravity. This gravitational water drains through the soil.

VADOSE ZONE

Vadose zone extends from the upper limit of the capillary zone to the lower edge of the soil-water zone as shown in Fig. 31.1. The thickness may vary from zero to more than 100 m. Zero thickness recorded, where the bounding zones merge with a high water table approaching ground surface and more than 100 m thickness under deep water table conditions. This zone is in other words called as *connecting zone* because it connects zone near the ground surface with that near the water table through which water moving vertically downward must pass. Due to hygroscopic and capillary forces, the water in vadose zone is non-moving, but on temporary basis excess water migrate downward, as it is due to gravitation so it is called as *gravitational water*.

Capillary Zone

The capillary zone is that zone which extend from the water table up to the limit of capillary rise of water. Which is generally derived from an equilibrium between surface tension of water and raised water weight. Thus.

Where r = Surface tension

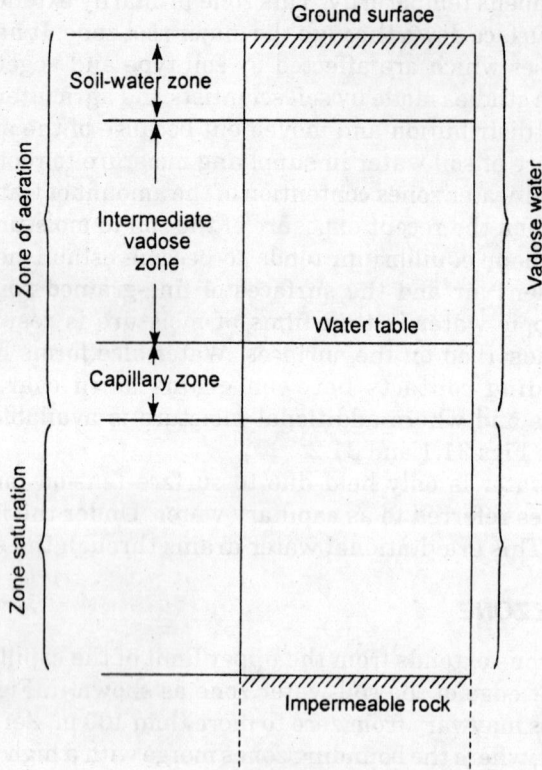

Fig. 31.1: *Divisions of subsurface water.*

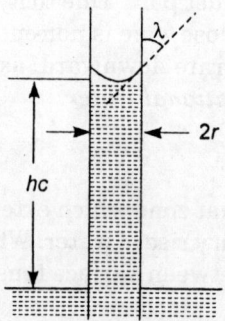

Fig. 31.2: *Rise of water in a capillary tube.*

y = Specific weight of water
r = Tube radius
h = Angle of contact between the meniscus and the well of the tube.

For pure water in clean glass: $h = 0$

$$A + 20° C \quad T = 0.074 \text{ g/cm}$$

$$Y = 1 \text{ g/cm}^3$$

So that the capillary rise approx. $hc = 0.15/r$.

We know from the above equation, the thikness of the capillary zone generally varies inversely with the pore size of a soil or rock capillary rise measurement in unconsolidated materials as shown in following table.

Capillary rise in unconsolidated materials

Material	Grain size (mm)	Capillary rise (cm)
Fine grarels	5–2	2.5
Very coarse sand	2–1	6.5
Coarse sand	1–0.5	13.5
Medium sand	0.5–0.2	24.6
Fine sand	0.2–0.1	42.6
Silt	0.1–0.05	105.5
Silt	0.05–0.02	200

We found that material containing innumerable pores of a wide range in size, when we study microscopically, a gradual decrease in water content results with height the upper boundary of the zone will form a jagged limit. Just above the water table almost all pores contain capillary water, as water decrease in pore spaces as further increase in height from water table. This distributions of water above the water table as in Fig. 31.2 which shows the visual capillary rise is invariably loss than actual capillary rise is invariably loss than actual capillary zone.

WATER CONTENT MEASUREMENT

As we know that the water content of soil determination can be accomplished by various direct methods based on removel of the water from a sample by evaporation, leaching, or chemical reaction which are followed by measurement of the amount removed. For which the gravimetric method involves weighing a wet soil sample, removing the water by over-drying it, and reweighing the sample.

Soil moisture is also measured by lowering neutron probe in a small-diameter tube in the ground, determination of soil moisture can be made as a function of depth. The instrument contains a radium-beryllium source of fast neutrons and a detector for show neutrons. The fast neutrons are slowed by collisions with hydrogen, and because most of the hydrogen in soil is associated with water, the intensity of slow neutrons measure the local soil moisture content.

We mark a negative-pressure head of water, which exists within the vadose zone, often referred to as suction or tension. This tension can be measured by a tensiometer.

Figure 31.3 shows a tensiometer installed in a soil column. The depression Δh in water level measures the local soil tension. Such

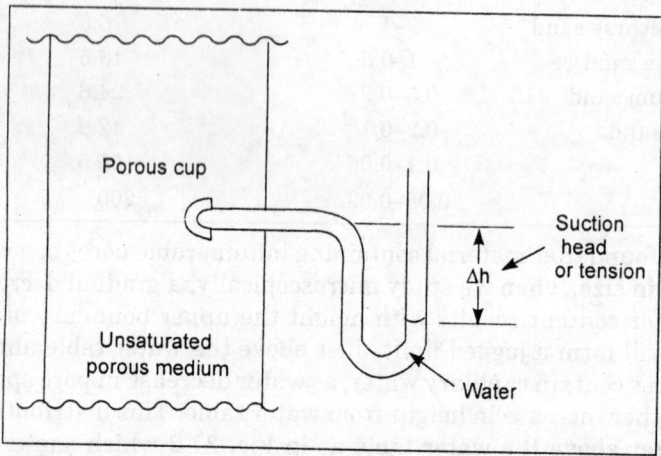

Fig. 31.3: *Illustration of a tensiometer for measuring water tension in unsaturated porous media.*

instruments function in the range from atmospheric pressure (near 1000 cm of water) to about 200 cm of wtaer (800 cm water tension). Calibration data reveal that the relation between the soil suction and water content is not single valued, instead influenced by the effects of wetting or drying.

AVAILABLE WATER IN SOILS

The most known characteristic of soils is that soils absorb and retain water, which may be withdrawn by plants during periods between rainfalls or irrigations. This water-holding capacity is defined by the plant available water, the moist end being the field capacity and the dry end the wilting point. Field capacity can be defined as the amount of water held in a soil after wetting and after subsequent drainage has become negligibly small. The negligible drainage rate is often assumed after two days. The wilting point deafness the water content of soils when plants growing in that soil are water content of soils when plants growing in that soil are reduced to a permanent witted condition. As soil type and volume and plant type and age influence witting point.

It is well-known fact that the ground water is safe water all purposes mainly drinking, industrial and agricultural generally known by the name of infiltration. The movement of water after entrance is known as *percolation*.

Porosity

Porosity is defined as the percentage of the voids present in given volume of soil/aggregate.
Mathematically,

$$\text{Porosity (P)} + = \frac{\text{Total volume of voids in soil/aggregate}}{\text{Total volume of soil/aggregate}}$$

The porosity of the unconsolidated material may be as high as 50%. The porosity of aggregates and sand is generally 4%. Commonly the porosity below 5% is called *small*, between 5 to 20% as *medium* and above 20% as *high*.

Aquiclude

The percolating water passes downward untill it reaches an impervious stratum known as aquiclude.

The ability of water-bearing stratum to store the water depends on its porosity and the size of particles. It the pores are small the resistance to water movement will be very great and it will be difficult to collect water in a well. Sands and gravels are the most important aquifer for public water supplies. Clay is highly porous but it has so fine particles that practically it is impervious.

In fact, limestone is less permeable and can allow movement of water only if it has caverns, cracks, faults or fissures.

(a) If possible the sediment should be kept in suspended stage in the reservoir, by means of mechanical stirrers. The suspended particles will flow out through sluices.

(b) We should grow vegetables, trees, grass, etc. which bind the soil particles of the ground, thus preventing erosion.

(c) As the sediment content increases just after the flood, this water should not be collected in the reservoir.

(d) Dam site should be selected at less erodable catchment area.

(e) Dam construction should be in stages.

(f) The vegetation screens should be provided at suitable places to check the entry of the suspended particles in the reservoir.

(g) Under sluices should be constructed in the dam to removed the silt of the reservoir.

 Definition of Aquifer Parameters

Aquifer

Geological formation and structure that transmit water in sufficient quantity to pumping well or springs is known as *aquifer*. In other words, a geological unit from which economic quantities of water can be extracted.

Confined Aquifer

A geological unit which contains water from which sufficient quantity of water may be extracted and are bounded above and below by an impermeable geological layer. The water in a confined aquifer is under pressure. In it well will rise up above the top of the aquifer. This level does not represent the water level in the aquifer, but is a measure of the pressure in the aquifer. This level is called the *head* (Fig. 32.1).

Unconfined Aquifer

An unconfined aquifer is generally called that aquifer in which a water table serves as the upper surface of zone of saturation.

Unconfined aquifer's base is an impermeable geological layer, but there is no impermeable layer, above.

It upper boundary is the water table.

If a well is constructed in the aquifer, the water level will be the same as the level of the water table. This water level is also called the *head*.

Confined and unconfined aquifer can be well understood by Fig. 32.2.

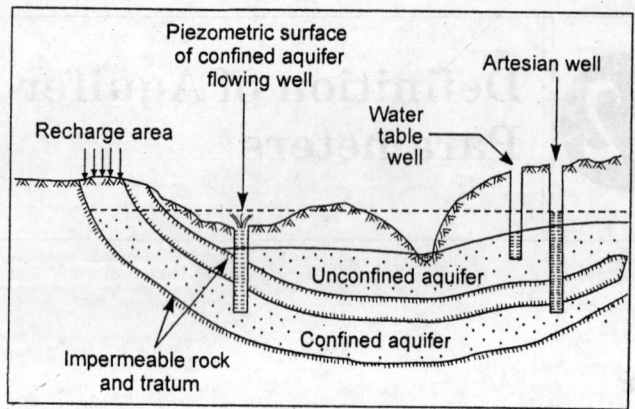

Fig. 32.1: *Ground water in confined and unconfined aquifers.*

WATER LEVEL ELEVATION AND GROUND WATER HEAD

Two different wells have different levels, to compare this, the water level must be measured from the same reference level. This is usually the mean sea level. The height above mean sea level is called the *elevation*.

When the water level in a well is measured, it is generally measured as a depth from the top of the well. It is generally converted into an elevation or ground water head, the elevation of the well must first be determined by surveyors.

As, then (Elevation of top of well) – (Measured depth to water).
= Ground water head

Storage Coefficient

Storage coefficient is nothing but a measure of how much water an aquifer can release. It has no unit because it is dimensionless as shown in Fig. 32.3.

Confined Storage Coefficient

Simply we can understand in this way that when water is pumped from a confined aquifer, the pressure in the aquifer reduces. This is generally shown by a fall in the head in wells in the aquifer.

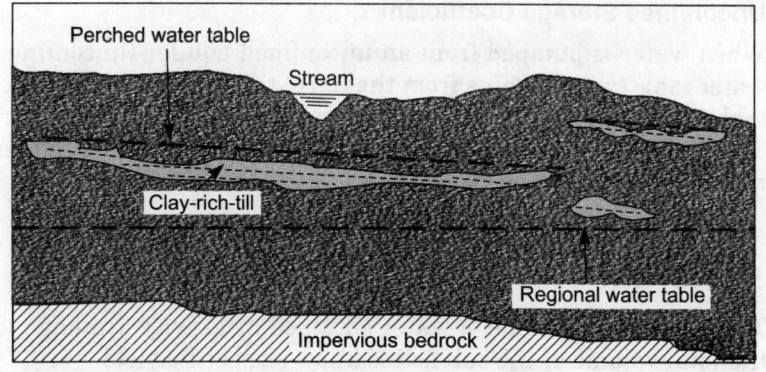

Fig. 32.2: *Perched water table supported by stringers of clay-rich till.*

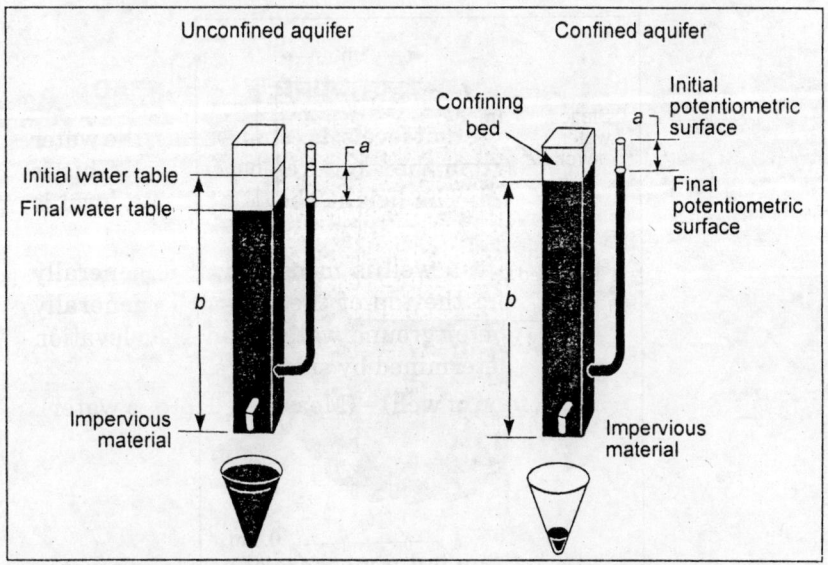

Fig. 32.3: *Unit prisms of unconfined and confined aquifers illustrating differences in storage coefficients. For equal declines in head, the yield from an unconfined aquifer is much greater than that from a confined aquifer. (After Heath and Trainer, 1968).*

The water abstracted comes from the expansion of the water and compaction of the aquifer due to the falling pressure.

Typical range is 0.00005 to 0.0005.

Unconfined Storage Coefficient

When water is pumped from an unconfined aquifer (unconfined water table) water drains from the top of the aquifer at the water table. The water abstracted comes from this draining of the aquifer. The storage coefficient of an unconfined aquifer is much higher than for a confined aquifer.

Typical range is 0.01 to 0.30.

Specific Yield

The ratio of the volume of water obtained by gravity-drainage to the total volume of the sub-soil is know as specific yield.

Specific yield = always less than porosity as shown in Fig. 32.4.

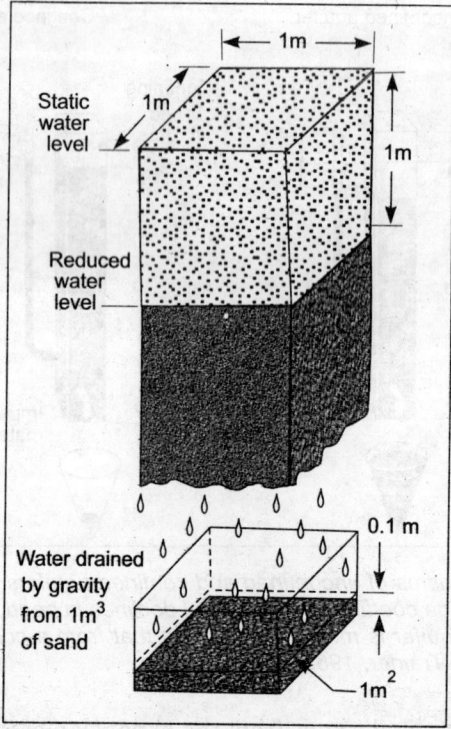

Fig. 32.4: *Specific yield of sand can be visualized from this diagram. Its value here is 0.1m³ per m³ of aquifer material.*

Specific Retention or Field Capacity

The ratio of the quantity of the water retained by the material against drainage to the total volume of the material or sub-soil is known as the *specipic retension* or *field capacity* of the soil.

Porosity

Specfic yield + specific retention when the size of the soil particles increases, the specific retention power decreases, whereas the specific yield increases.

Permeability: Permeability is the capability of a formation of soil to pass water through it. It is measured by the coefficient of permeability (K). Generally, permeability is defined as the rate of flow of water through an aquifer of unit cross-sectional area, under unit hydraulic gradient at 15°C temperature.

In other words, the permeability of a rock or soil defines its ability to transmit a fluid. If a rock has a high permeability, water moves through it easily. In hydrogeology, this is also called the *hydraulic conductivity*. Its units are metres/day.

Permeability coefficients

Type of soil	Coefficient of permeability in cm / sec
Clean gravel	1.0 and above
Coarse sand (clean)	1.0 to 1.01
Medium sand	0.01 to 0.005
Fine sand	0.005 to 0.001
Sand mixed with silt	2×10^{-2} to 1.0×10^{-4}
Silt	5×10^{-7} to 1.0×10^{-5}
Clay	1.0×10^{-6} and less.

The subsurface water velocity through aquifer, depends on various factors, such as type of soil, gradient, viscosity of water, arrangement of soil grains, porosity of soils, etc. describe that the flow of ground water will be laminar or turbulent.

Darcy's Law

The velocity of the water through the porous medium is inversely proportional to the length of the flow path and directly laminar range is proportional to the head loss or hydraulic gradient (R). As the head loss is also known as *potential loss*.

$Q \propto s$

$Q = KSA$

Or, $Q/A = KS = V$ (say)

Where, K = Darcy's coefficient of permeability of the soil

V = Velocity of the underground water.

A = Total area

S = Slope of the hydraulic gradient line.

From the above, it can also be assumed as,

$$\frac{A_v}{A} = \frac{V_v}{V} = P$$

Where, A_v = Area of the voids through which water percolates

V_v = Actual velocity of the percolation

P = Porosity of the soil

Now $Q = A_v V_v = P A_v V_v$

But as, $Q = AV$

Therefore, $A_v V_v = P A_v V_v$

Or $V = P V_v$

$\therefore$ $V_v = \dfrac{V}{P}$

Thus, we have seen above that the actual velocity of percolation can be determined by dividing the value of V (as obtained by Darcy law) by the porosity of the medium.

In this modern age Darcy's law is considered as fundamental equation in hydrology. It is generally used to calculate how much water is flowing through an aquifer.

$Q = KAi$

Where Q is the flow rate (m³/d)

K is the permeability (m/d)

A is the cross-sectional area of flow (m²)

i is the hydraulic gradient (dimension less)

The cross-sectional area, $A = DW$, where D is aquifer thickness and W is the width of flow.

Therefore, Darcy's law can also be written as

$Q = KDWi$

And since KD = transmissivity, T

$Q = TWi$

Transmissivity (T)

Hydrogeologists generally use the term transmissivity instead of permeability transmissivity is the permeability multiplied by the vertical thickness of the equifer, D.

$$T = KD$$

In a confined aquifer, the saturated thickness, and therefore, the transmissivity remains constant.

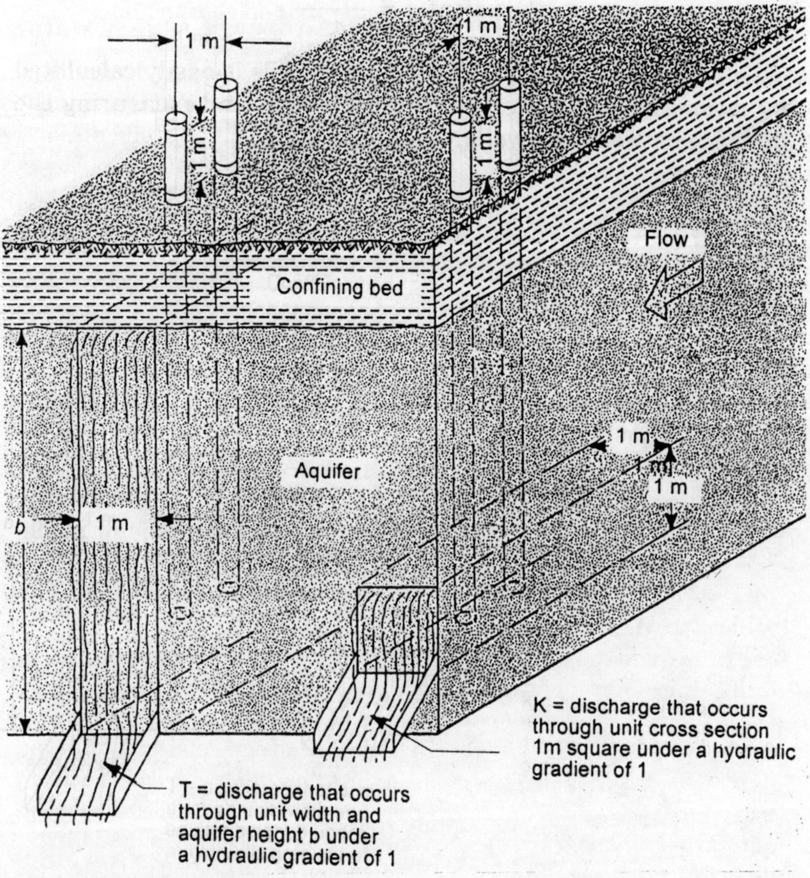

Confining bed

Flow

Aquifer

1 m

1 m

1 m

1 m

1 m

b

1 m

K = discharge that occurs through unit cross section 1m square under a hydraulic gradient of 1

T = discharge that occurs through unit width and aquifer height b under a hydraulic gradient of 1

Fig. 32.5: *Coefficients of hydraulic conductivity and transmissivity. Hydraulic conductivity multiplied by the aquifer thickness equals coefficient of transmissivity.*

In an unconfined aquifer, the transmissivity and the saturated thickness vary as the water level changes.

Units are m^2/d as shown in Fig. 32.5.

Hydraulic Gradient (i)

The hydraulic gradient is the difference in ground water head between two points (on a flow line) devided by the distance between them.

$$\text{Gradient, } i = \frac{h_2 - h_1}{x}$$

The hydraulic gradient between two wells is easily calculated by measuring the water level in each well, and measuring the distance between the wells.

Ground Water Levels and Environmental Influences

In general "ground water" or "subsurface water" is that water which occurs below the surface of the earth and main source of it is infiltration. A ground water level indicates the elevation of atmospheric pressure of the aquifer. Any phenomenon that produces a change in pressure on ground water will cause the ground water level to vary. Differences between supply and withdrawal of ground water cause levels to fluctuate. Stream flow variations are closely related to ground water levels. Other causes of ground water levels fluctuation include meteorological and tidal phenomena, urbanization, earthquakes and external loads. And, last but not least subsidence of the land surface can occur due to changes in underlying ground water conditions.

GROUND WATER LEVEL VARIATIONS

Secular Variations

Alternating series of wet and dry years, in which the rainfall is above or below the mean, produce long-period fluctuations of levels. The long records of rainfall and grundwater levels as shown in Fig. 33.1. Rainfall is not an accurate indicator of ground water level changes but recharge which is the governing factor mainly depends on rainfall intensity and distribution and amount of surface runoff.

In overdeveloped basins where draft exceeds recharge, a downward trend of ground water levels may continue for many years. Figure 33.2 shows the decline in piezometric surface of a deep sandstone aquifer as a result of nearly a century of intensive pumping in the Chicago metropolitan area.

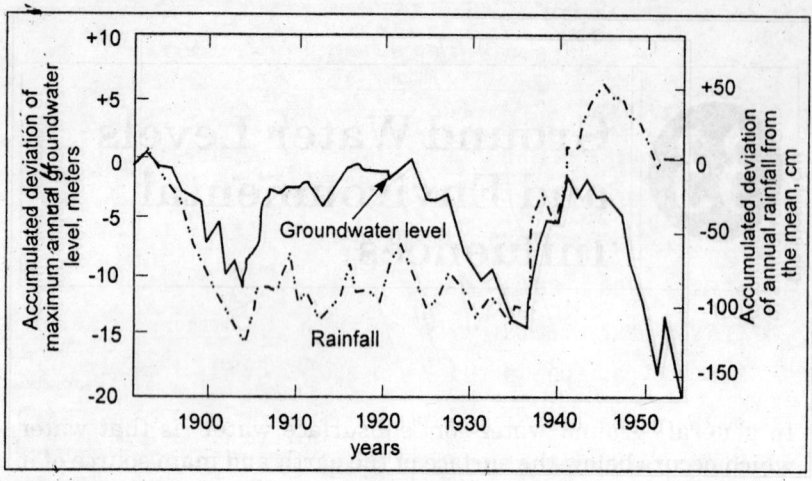

Fig. 33.1: *Secular variations of maximum annual ground water level and annual rainfall.*

Seasonal Variations of Ground Water Levels

Seasonal pattern of fluctuation is due to influences such as rainfall and irrigation pumping that follow well-defined seasonal cycles. Highest levels occur in late spring and lowest in winter. In irrigated areas where frozen ground is not a factor, lowest levels normally occur during fall at the end of the irrigation season. The amplitude depends on recharge, pumpage, and the type of aquifer. It is interesting to note that confined aquifers normally display a greater range in levels than do unconfined aquifers.

Another type of ground water level variation may be **short-term variation** in which ground water levels often display characteristic short-term fluctuations governed by the primary use of ground water in a locality. Clearly defined diurnal variations may be associated with municipal water-supply wells and weekly patterns occur with pumping for industrial and muncipal purposes.

STREAM FLOW AND GROUND WATER LEVELS

Depending on the relative levels, if a stream channel is in direct contact with an unconfined aquifer, the stream may recharge the ground water or receive discharge from the ground water, Generally, we name gaining stream is one receiving ground water

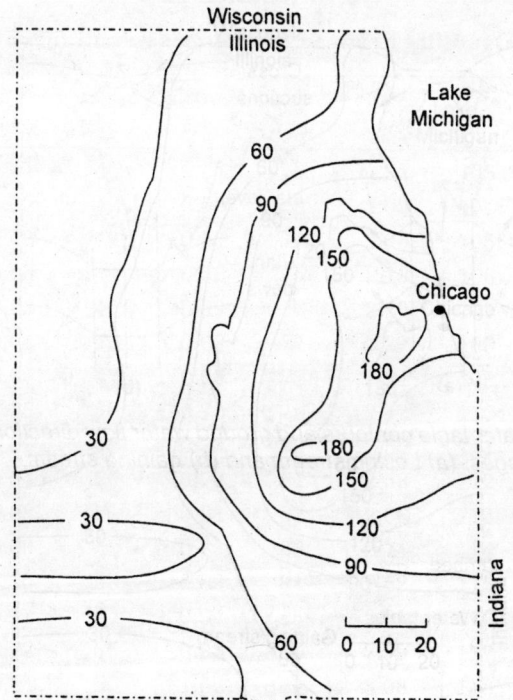

Fig. 33.2: *Decline of the piezometric surface in the Chicago metropolitan area due to extended heavy pumping. Contours are lines of equal decline in metres.*

discharge while a losing stream is one recharging ground water (Fig. 33.3).

The rising water is that water which marked increases in stream flow in reaches where a subsurface restriction forces ground water to the surface (Fig. 33.4).

Bank Storage

During flooded stream period ground water levels are temporarily raised near the channel by inflow from the stream. The volume of water so stored and released after the flood is referred to as bank storage to evaluate bank storage and its rate of inflow and outflow; analytic or model approaches are necessary to obtain quantitative estimates for specified boundary conditions because field data are rarely adequate.

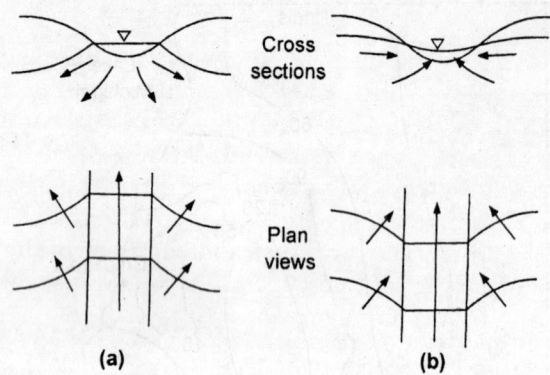

Fig. 33.3: *Water table contours and ground water flow directions in relation to stream stages. (a) Losing stream and (b) gaining stream.*

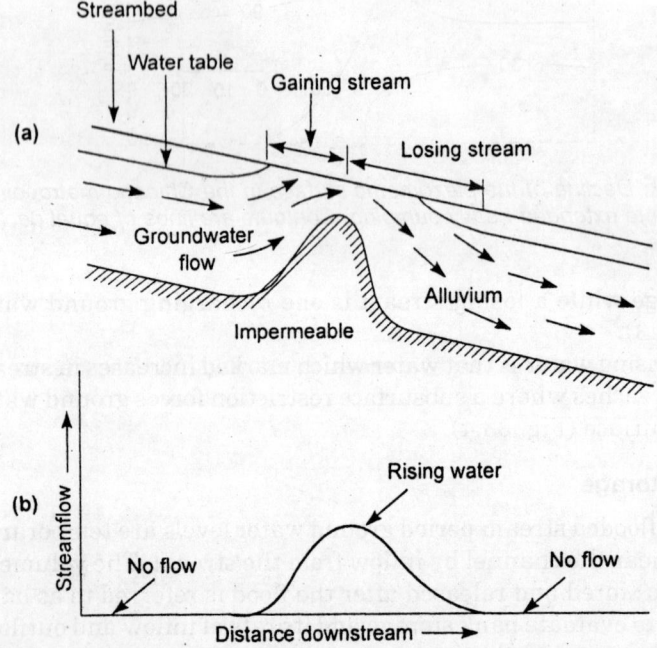

Fig. 33.4: *Rising water in a stream channel from emerging ground water flow. (a) Cross-section along stream channel in an alluvial valley and (b) stream flow as a function of distance along the stream.*

Two renowned scientists, Cooper and Rorabaugh, derived solutions for changes in ground water head near the stream, ground water flow to the stream, and bank storage. Their comprehensive analysis also included a family of asymmetric flood-wave stage hydrographs, which facilitate study of the effects of a wide variety of flood shapes on ground water.

Bank storage study produced by the spring rise along an 80 km indicates a total storage volume (including both banks of the stream) of some 2.07×10^8 m^3. During a typical 45 day rise of the river to flood peak, bank storage of 4.60×10^6 m^3/day in the subsequent 165 day decline, the return flow averages 1.25×10^6 m^3/day.

Base Flow

Base flow or ground water run-off is referred as stream flow originating from ground water discharge. During precipitation periods stream flow is derived primarily from surface run-off, while during dry periods all stream flow may be contributed by base flow. Typically, base flow is not subject to wide fluctuations and is indicative of aquifer charcteristics within a basin.

Base flow estimation is done by a rating curve of ground water run-off, which can be prepared by plotting mean ground water stage (water table level) within the basin against stream flow during periods when all flow is contributed by ground water. It may be noted that frozen ground impeded ground water recharge during February and March and that base flows are largest during the spring and summer months. Ground water contributed 33 percent of total stream flow for the year.

Stream flow at any instant contains ground water contributed at previous times and different locations within the drainage area. During and after a storm period in a small drainage basin, the water table will rise, causing the base flow to increase also. But superimposed on this will be the bank storage fluctuation (Fig. 33.5). The effects of these two variations are shown schematically in Fig. 33.6.

To determine the separation of total stream flow into surface run-off and grundwater components during flood periods can be accomplished from measurements of chemical concetrations. Total dissolved solids or any major ion will serve the purpose with the equation $C_{TR} Q_{TR} = C_{CW} Q_{CW} + C_{SR} Q_{SR}$ where C is ionic concentration, Q is stream flow, TR is total run-off, GW is ground

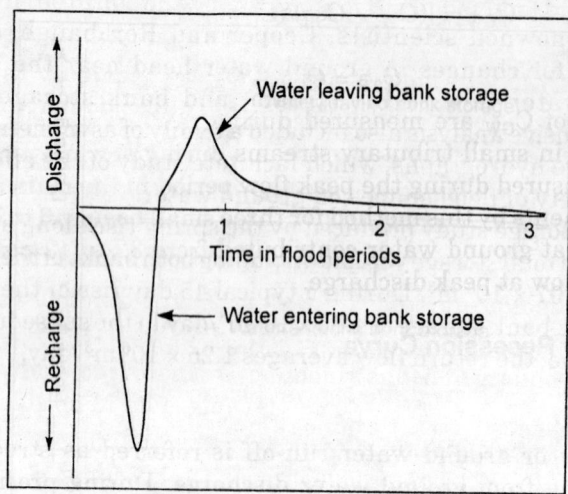

Fig. 33.5: *Ground water flow to and from bank storage.*

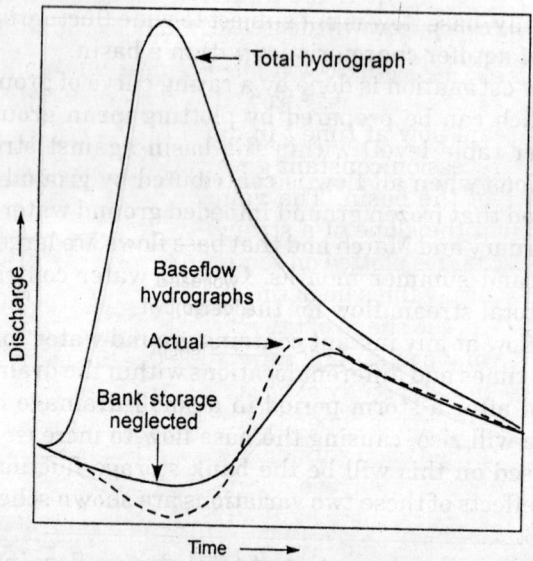

Fig. 33.6: *Schematic diagram of the variation of base flow during a flood hydrograph with and without effects of bank storage (after Singh[62]).*

water contribution (base flow), and SR is surface run-off. For the base flow,

$$Q_{GW} = [(C_{TR} - C_{SR})/(C_{GW} - C_{SR})] \, Q_{TR}$$

where

$$Q_{TR} = Q_{GW} - Q_{SR}$$

Values of C_{GW} are measured during rainless periods, C_{SR} is measured in small tributary streams during storm events, and C_{TR} is measured during the peak flow period in the main stream. Measurements by this method for three small basins (6 to 13 km^2) showed that ground water contributed from 32 to 42 percent of the total flow at peak discharge.

Base Flow Recession Curve

A base flow recession curve shows the variation of base flow with time during periods of little or no rainfall over a drainage basin. In other words it is a measure of the drainage rate of ground water storage from the basin. In large highly permeable aquifers within a drainage area, the base flow will be sustained even through prolonged droughts; but if the aquifers are small and of low permeability, the base flow will decrease relatively rapidly and may even cease the recession curve can often be fitted by the equation

$$Q = Q_0 K^t$$

where Q is stream flow at time t in days after a given discharge Q_0, and K is a recession constant governed by the hydrogeologic characteristics of the basin. The value of K can be empirically determined from the slope of a straight line fitted to a series of consecutive discharges plotted on semilogarithmic paper, as shown in Fig. 33.7. Typical values lie in the range of 0.89 and 0.95. Thus, prior knowledge of the shape of the recession curve enables future estimates to be made of stream flow during rainless periods.

Singh made analytic study of base flow and demonstrated that base flow recession curves depend on the degree to which a stream channel is entrenched in an aquifer. For a fully penetrating stream (Fig. 33.8 a), recession curves do not plot as straight lines on semilogarithmic paper; instead, the recession rate continuously decreases with time, forming a concave curve. But for deep aquifers and partially penetrating streams (Fig. 33.8 b), the straight-line approximation is generally applicable. The value of K in Eq. $Q = Q_0 K^t$ varies directly with the degree of stream entrenchment.

These approaches to base flow assume that ground water drains only toward the stream channel. Ground water also can flow

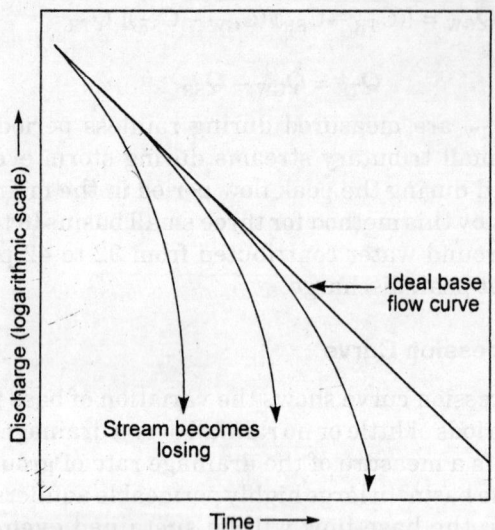

Fig. 33.7: *Base flow recession curves of stream flow for varying magnitudes of evapotranspiration losses from ground water.*

downward to an underlying leaky aquifer and can be lost by evapotranspiration to the atmosphere. Where these diversions are significant, the recession curve will be deflected downward. In semiarid regions where stream flow is intermittent, evapotranspiration losses become significant; this causes the recession curve to steepen (Fig. 33.9) until stream flow finally ceases.

FLUCTUATIONS OF GROUND WATER LEVELS DUE TO EVAPOTRANSPIRATION

Evapotranspiration is the sum of water used by plants in a given area in transpiration and the water evaporated from the adjacent soil in the area in any specified time. Unconfined aquifers with water tables near ground surface frequently exhibit diurnal fluctuations that can be ascribed to evaporation and transpiration. Both of them (processes) cause a discharge of ground water into the atmosphere and have nearly the same diurnal variation because of their high correlation with temperature.

Evaporation Effects

Evaporation from ground water increases as the water table approaches ground surface. The rate also depends on the soil

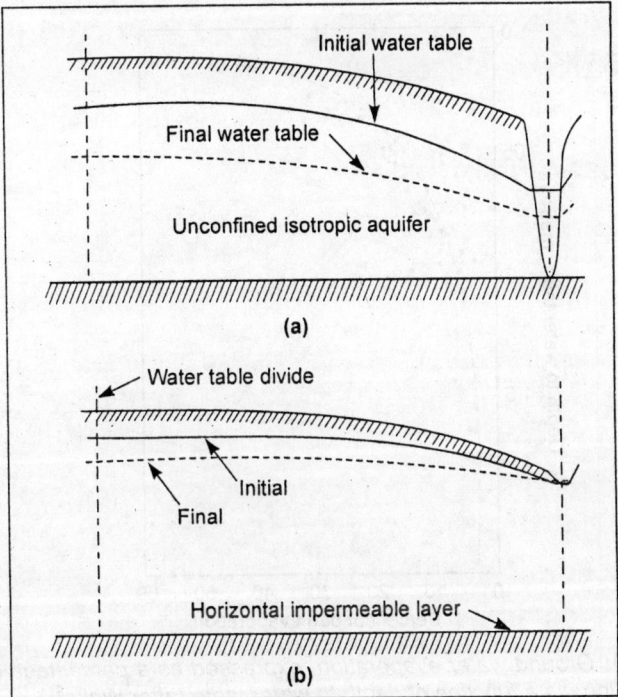

Fig. 33.8: *Water table and stream channel conditions affecting base flow. (a) Fully penetrating stream and (b) partially penetrating stream (after Singh[62]).*

structure, which controls the capillary tension above the water table and hence its hydraulic conductivity. For isothermal conditions, upward movement is essentially all in the liquid phase, but a soil may have a high surface temperature, causing it to dry out and establishing upward vapour movement in response to a vapour pressure gradient.

Field measurements of ground water evaporation from tanks filled with soil (lysimeters) have been made and water tables were maintained at prescribed depths below ground surface. Which resulted as and expressed as a percentage of pan evaporation at ground surface, are shown in Fig. 33.9. For water tables within one metre of ground surface, evaporation is largely controlled by atmospheric conditions, but below this soil properties become limiting and the rate decreases markedly with depth.

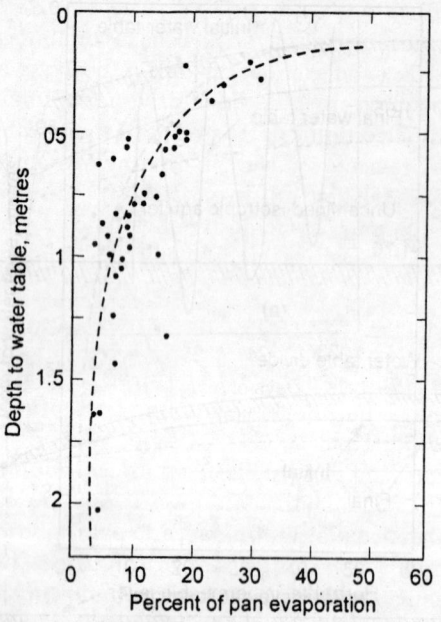

Fig. 33.9: *Ground water evaporation, expressed as a percentage of pan evaporation, as a function of depth to water table (after White[81]).*

Transpiration Effects

When the root zone of vegetation reaches the saturated stratum, the uptake of water by roots equals the transpiration rate. Figure 33.10 shows water level variations measured in a well in a thickest of willows. Rapid foliage growth during August (Fig. 33.10 a) caused daily fluctuation averaging about 10 cm with the water table between 1.6 and 1.8 m below ground surface. Due to occurance of heavy frost in early October and most leaves fall by mid-October; thereafter, diurnal fluctuations are negligible (Fig. 33.10 b) with the vegetation dormant.

In fact, magnitudes of transpiration fluctuations depend on the type of vegetation, season, and weather. Hot, windy days produce maximum drawdowns, whereas cool, cloudy days show only small variations. Fluctuations begin with the appearance of foliage and cease after killing frosts. Transpiration discharge does not occur in nonvegetated areas, such as ploughed fields, or in areas where the water table is far below ground surface. After rain on high

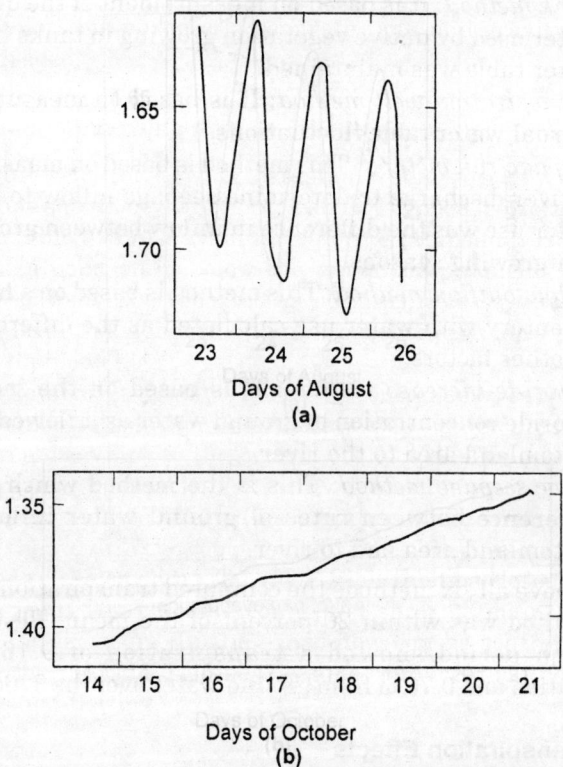

Fig. 33.10: *Effect of transpiration discharge on ground water levels near Milford, Utah. (a) In summer and (b) after frost (after White[81]).*

water table vegetated land, the water table rises sharply as the increased soil moisture meets the transpiration demand and reduces the ground water discharge; but on cleared land or when vegetation is dormant, little or no rise is evident.

In an interesting field study. Lewis and Burgy demonstrated that 6–12 m oak trees in California extracted water from a water table in fractured and jointed rock at a depth of 21 m. Tritium injected in a nearby well was found subsequently in water transpired by the trees (it was made by collecting from plastic bags tied over ends of branches).

The transpiration of ground water by phreatophytes was studied by the US Geological Survey in Lower Safford Valley, Arizona. Six methods for determing water use by vegetation were employed.

1. *Tank method*: It is based on measurement of the quantity of water used by native vegetation growing in tanks in which a water table was maintained.
2. *Transpiration-well method*: It is based on measurement of diurnal water table fluctuations.
3. *Seepage-run method*: This method is based on measurements of river discharge to determine seepage inflow to the river; water use was the difference in inflow between growing and non-growing seasons.
4. *Inflow-outflow method*: This method is based on a hydrologic inventory with water use calculated as the difference from all other factors.
5. *Chloride-increase method*: It is based on the increase in chloride concentration of ground water as it flowed from the bottomland area to the river.
6. *Slope-seepage method*: This is the method which based on difference between rates of ground water inflow to the bottomland area and to river.

For above all six methods the computed transpiration value for each method was within 20 percent of the mean. Results for a 12 month period showed a transpiration of 0.16 m from precipitation and 0.75 m from ground water over the 3765 ha area.

Evapotranspiration Effects

The combined loss of ground water referred to as evapotranspiration. The variation of evapotranspiration with water table depth as Fig. 33.11 for three groundcover conditions. It is apparent that the deeper the roots, the greater the depth at which water losses occur. Even with relatively deep water tables, evapotranspiration does not necessarily become zero because upward transport can still occur, albeit minimally, in the vapour phase.

In general, the maximum water table level occurs in mid-morning (see Fig. 33.12) and represents a temporary equilibrium between discharge and recharge from surrounding ground water. From midmorning until early evening, losses exceed recharge and the level falls. The steep slope near mid-day indicates maximum discharge associated with highest temperatures. The evening minimum again represents an equilibrium point, while the rise during the night hours is recharge in excess of discharge.

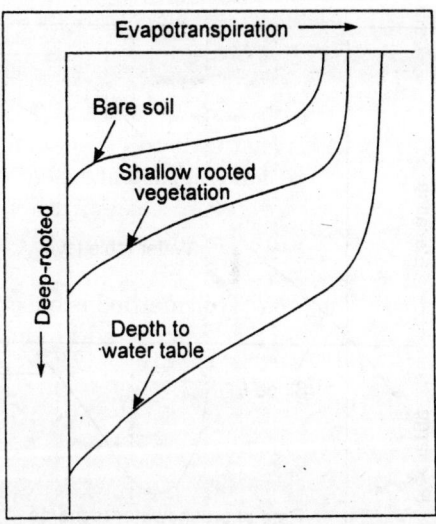

Fig. 33.11: *Generalised variation of evapotranspiration from ground water with water table depth for three ground-cover conditions (after Bouwer[3]).*

A method for computing the total quantity of ground water withdrawn by evapotranspiration during a day. It is assumed that evapotranspiration is negligible from midnight to 4 AM and further, that the water table level during this interval approximates the mean for the day, then the hourly recharge from midnight to 4 AM may be taken as the average rate for the day. Suppose h equal the hourly rate of the water table from midnight to 4 AM, as shown by the upper curve in Fig. 33.12, and s the net fall or rise of the water table during the 24-hour period, then as a good approximation the diurnal volume of ground water discharge per unit area

$$\overline{V}_{ET} = S_V \, (24h \pm s)$$

where S is the specific yield near the water table. Troxell, pointed out the rate of ground water recharge to the vegetated area varies inversely with the table level. Thus, evapotranspiration from ground water can serve to stabilize the water table near ground surface.

"It is also important to point out that in areas where evapotranspiration serves to control the position of the water table, waterlogging and salination of soils in the root zone occur. Agricultural

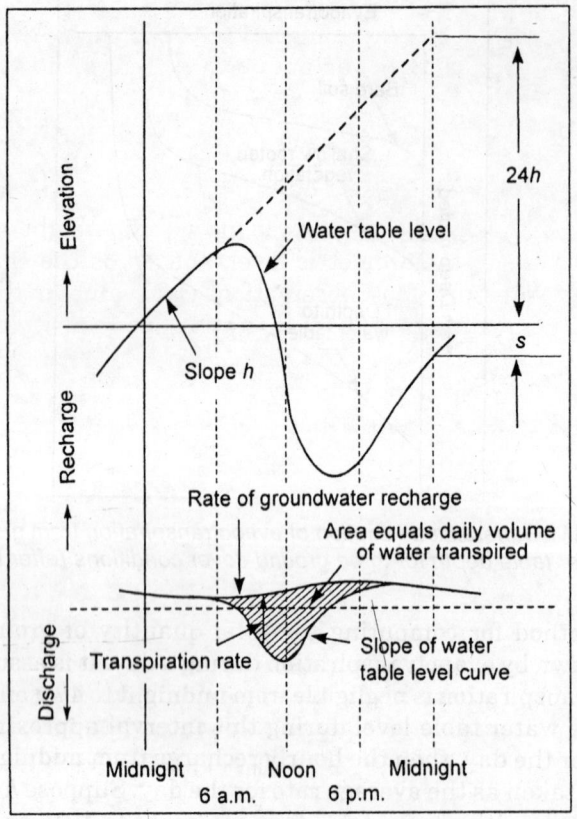

Fig. 33.12: *Inter-relations of water table level, recharge, and evapotranspiration fluctuations (after Troxell[68]).*

areas so afflicted can become saline wastelands unless adequate drainage systems are installed."

FLUCTUATIONS DUE TO METEOROLOGICAL PHENOMENA
Atmospheric Pressure

Atmospheric pressure changes make change in ground water level that means changes in atomspheric pressure which produce sizable fluctuations in wells penetrating confined aquifers. The relationship is inverse, i.e. increases in atmospheric pressure produce decreases in water the ratio of water levels, and conversely. When atmospheric pressure changes are expressed in terms of a column of water, level change to pressure change expresses the barometric efficiency of an aquifer which is as shown in this equation:

$$B = \frac{\gamma \Delta h}{\Delta p_a}$$

where B is barometric efficiency, γ is the specific weight of water, Δh is the change in piezometric level, and Δp_a is the change in atmospheric pressure. Most observations yield values in the range of 20 to 70 percent. Figure 33.13 shows the upper curve indicates observed water levels in a well penetrating a confined aquifer. The lower curve shows atmospheric pressure inverted, expressed in meters of water, and multiplied by 0.75. A close correspondence of major fluctuations exists in the two curves; the equality of amplitudes indicates that the barometric efficiency of the aquifer is about 75 percent.

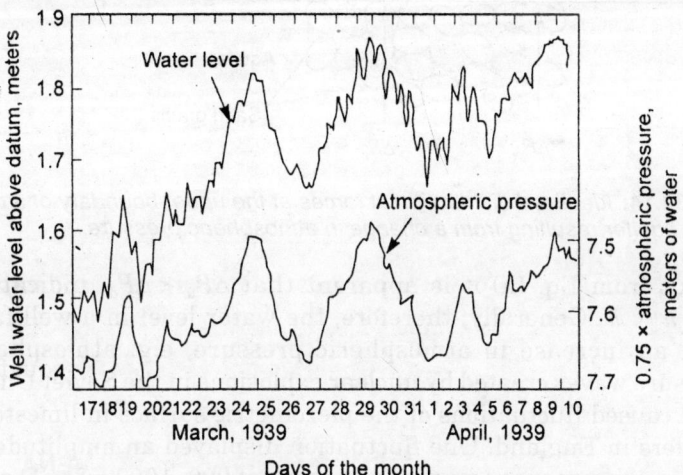

Fig. 33.13: *Response of water level in a well penetrating a confined aquifer to atmospheric pressure changes, showing a barometric efficiency of 75 percent (after Robinson[58]).*

According to equation as if Δp_a is the change in atmospheric pressure and Δp_w is the resulting change in hydrostatic pressure at the top on a confined aquifer, then $\Delta P_a = \Delta P_w + \Delta S_c$ (a) where

ΔS_c is the increased compressive stress on the aquifer (Fig. 33.14). At a well penetrating the confined aquifer, the relation

$$P_w = P_a + \gamma h \qquad\qquad \dots \text{(b)}$$

exists as shown in Fig. 33.14, where γ is the specific weight of water. Let the atmospheric pressure increase by Δp_a, then

$$P_w + \Delta P_w = P_a + \Delta P_a + \gamma h \qquad\qquad \dots \text{(c)}$$

as shown in Fig. 33.14. Substituting for P_w from Eq. (b) yields

$$\Delta P_w = \Delta P_a + \gamma (h - h) \qquad\qquad \dots \text{(d)}$$

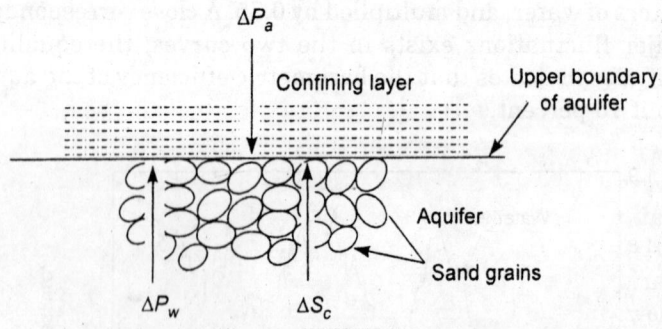

Fig. 33.14: *Idealized distribution of forces at the upper boundary of a confined aquifer resulting from a change in atmospheric pressure.*

But from Eq. (a) it is apparent that $\Delta P_w < \Delta P_a$, indicating that $h' < h$. Generally, therefore, the water level in a well falls with an increase in atmospheric pressure, e.g. atmospheric pressure waves created by nuclear explosions in the Soviet Union have caused fluctuations of the piezometric surface in limestone aquifers in England. One fluctuation displayed an amplitude of 0.46 cm in response to a pressure wave of 900–1000 microbars.

A developed expressions by Jacob relating barometric efficiency of a confined aquifer to aquifer and water properties, including the storage coefficient. Gilliland showed that changes in soil moisture from infiltrating precipitation can affect the magnitude of barometric efficiency.

Atmospheric pressure changes are transmitted directly to the water table, both in the aquifer and in a well; hence, no pressure difference occurs. It is the case with an unconfined aquifer. Air entrapped in pores below the water table is affected by pressure changes, however, causing fluctuations similar but smaller than that observed in confined aquifers. Temperature fluctuations in the capillary zone will also induce water table fluctuations where entrapped air is present.

Fluctuations of atmospheric pressure do affect water tables substantially on small, permeable oceanic islands. The response of sea-level changes to atmospheric pressure is essentially isostatic, i.e. sea level adjusts to a constant mass of the ocean-atmosphere column. This causes the ocean to act as an inverted barometer with sea level rising about 1 cm to compensate for a drop in atmospheric pressure of 1 mb.

Rainfall

Generally, we consider that rainfall is not an accurate indicator of ground water recharge because of surface and subsurface losses as well as travel time for vertical percolation. The travel time may very from a few minutes for shallow water tables in permeable formations to several months or years for deep water tables underlying sediments with low vertical permeabilities. In arid and semiarid regions, recharge from rainfall may be essentially zero. Shallow water tables show definite responses to rainfall.

Droughts extending over a period of several years contribute to declining water levels. Where the unsaturated zone above a water table has a moisture content less than that of specific retention the water table will not respond to recharge from rainfall until this deficiency has been satisfied. Thereafter, the rise Δh will amount to

$$\Delta h = P_i / S_v$$

where P_i is that portion of precipitation that percolates to the water table and S_y is specific yield.

An interesting phenomenon occasionally noted in observation wells. If the zone containing interconnected air-filled pores (H in Fig. 33.15) is compressed to a thickness $H - m$, then the pressure above the water table is increased by $m/(H - m)$ of an atmosphere, causing the water level in an observation well to rise.

$$\Delta h = \frac{m}{H-m} \,(10)\, m$$

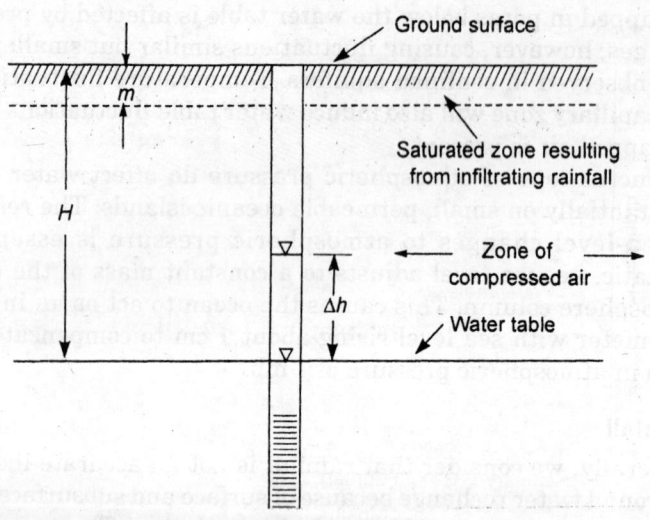

Fig. 33.15: *Water table rise in an observation well resulting from infiltrating rainfall sealing the ground surface and compressing air above the water table.*

Wind

Generally, when a gust of wind blows across the top of a casing, the air pressure within the well is suddenly lowered and, as a consequence, the water level quickly rises. After the gust passes, the air pressure in the well rises and the water level falls. Minor fluctuations of water levels are caused by wind blowing over the tops of wells. The effect is illustrated by Fig. 33.16.

Frost

In regions of heavy frost it has been observed that shallow water tables decline gradually during the winter and rise sharply in early spring before recharge from ground surface could occur as in Fig. 33.17. This fluctuation can be attributed to the presence of frost layer above the water table. During winter water moves upward from the water table by capillary movement and by vapour transfer to the frost layer where it freezes. Vapour migration occurs in response to the thermal gradient and to the fact that vapour pressure over ice is less than that over liquid water at 0°C. In early spring, approximately when the mean air tempe-

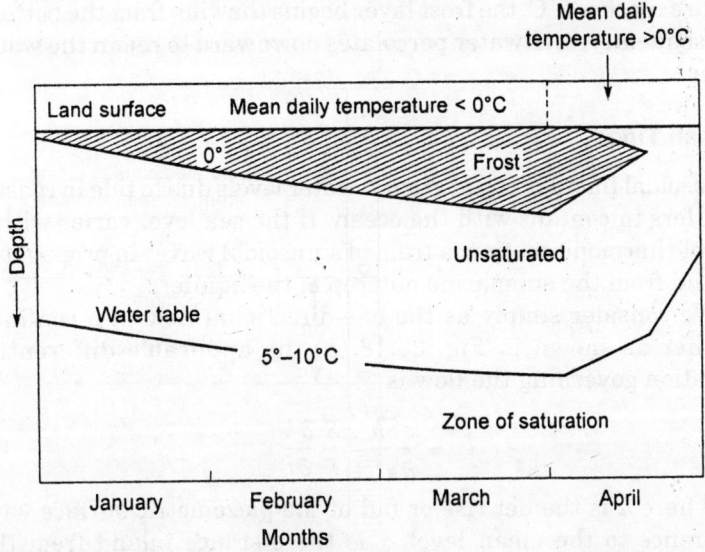

Fig. 33.16: *Wind-induced water level fluctuations in a well at Miami, Florida, during passage of a hurricane.*

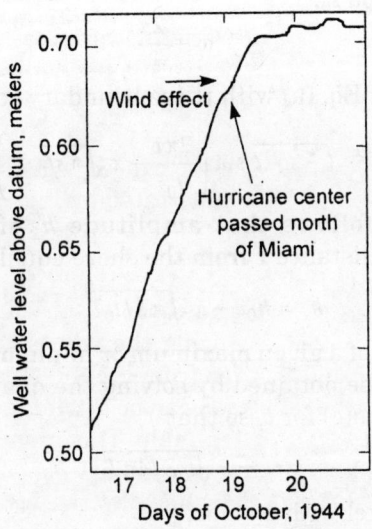

Fig. 33.17: *The variation in depth to water table in response to winter frost conditions.*

rature reaches 0°C, the frost layer begins thawing from the bottom; consequently, meltwater percolates downward to rejoin the water table.

Ocean Tides

Sinusoidal fluctuations of ground water levels due to tide in coastal aquifers in contant with the ocean. If the sea level varies with a simple harmonic motion, a train of sinusoidal waves in propagated inland from the submarine outcrop of the aquifer.

We consider simply as the one-directional flow in a confined aquifer as shown in Fig. 33.18 (a) the applicable differential equation governing the flow is

$$\frac{\partial^2 h}{\partial x^2} = \frac{S}{T}\frac{\partial h}{\partial t} \tag{a}$$

Where h is the net rise or fall of the piezometric surface with reference to the mean level, x is the distance inland from the outcrop, S is the storage coefficient of the aquifer, T is transmissivity, and t is time. Letting the amplitude, or half range, of the tide be h_0, $x = 0$ and $h = 0$ at $x = \infty$. The angular velocity is ω; for a tidal period t_0,

$$\omega = \frac{2\pi}{t_0} \tag{b}$$

The solution of Eq. (a) with these boundary conditions is

$$h = h_0 e^{-r}\sqrt{-s-t}\,\sin\left(\frac{2\pi t}{t_0} - x\sqrt{\pi s t_0\ T}\right) \tag{c}$$

From this it follows that amplitude h_x of ground water fluctuations at a distance x from the shore equals

$$h_x = h_0 e - x\sqrt{\pi s/t_0 T} \tag{d}$$

The time lag t_L of a given maximum or minimum after it occurs in the ocean can be obtained by solving the quantity within the parentheses of Eq. (c) for t, so that

$$t_L = x\sqrt{t_0 S/4\pi T} \tag{e}$$

The waves travel with a velocity

$$v_\omega = \frac{x}{t_L} = \sqrt{4\pi T/t_0 S} \tag{f}$$

and the wavelength is given by

$$L_\omega = v_\omega t_0 = \sqrt{4\pi t_0 T / S} \qquad \text{(g)}$$

Substituting the wavelength for x in Eq. (d) shows that the amplitude decreases by a factor e^{-2x}, or 1/535, for each wavelength. Water flows into the aquifer during half of each cycle and out during the other half. By Darcy's law the quantity of flow V per half-cycle per foot of coast is

$$V = \int_{-t_0/8}^{3\,t_0/8} q\,dt = T \int_{-t_0/8}^{3\,t_0/8} \left(\frac{\partial h}{\partial x} \right)_z dt \qquad \text{(h)}$$

Where q is the flow per foot of coast. Differential Eq. (c) to obtain $\partial h / \partial t$ and integrating yields

$$V = h_0 \sqrt{2 t_0 ST / \pi} \qquad \text{(i)}$$

The above said analysis is also applicable as a good approximation to water table fluctuations of an unconfined aquifer if the range of fluctuation is small in comparison to the saturated thickness (Fig. 33.18 b).

Variations of piezometric levels, occur due to change in atmospheric pressure, so do tidal fluctuations vary the load on confined aquifers extending under the ocen floor (Fig. 33.18 c). Contrary to the atmospheric pressure effect, tidal fluctuations are direct; that is, as the sea level increases, the ground water level also increase. The ratio of piezometric level amplitude to tidal amplitude is known as the *tidal efficiency* of the aquifer. Jacob showed that tidal efficiency C is related to barometric efficiency B by

$$C = 1 - B \qquad \text{(j)}$$

In other words tidal efficiency is a measure of the incompetence of overlying confining beds to resist pressure changes. Aquifer response to loading rather than head change at the outcrop requires that the amplitude given by Eq. (c) be multiplied by C.

Earth Tides

Regular semidiurnal fluctuations of small magnitude have been observed in piezometric surfaces of confined aquifers located at great distances from the ocean. Figure 33.18 shows fluctuations over a lunar cycle from a 250 m well tapping a confined aquifer.

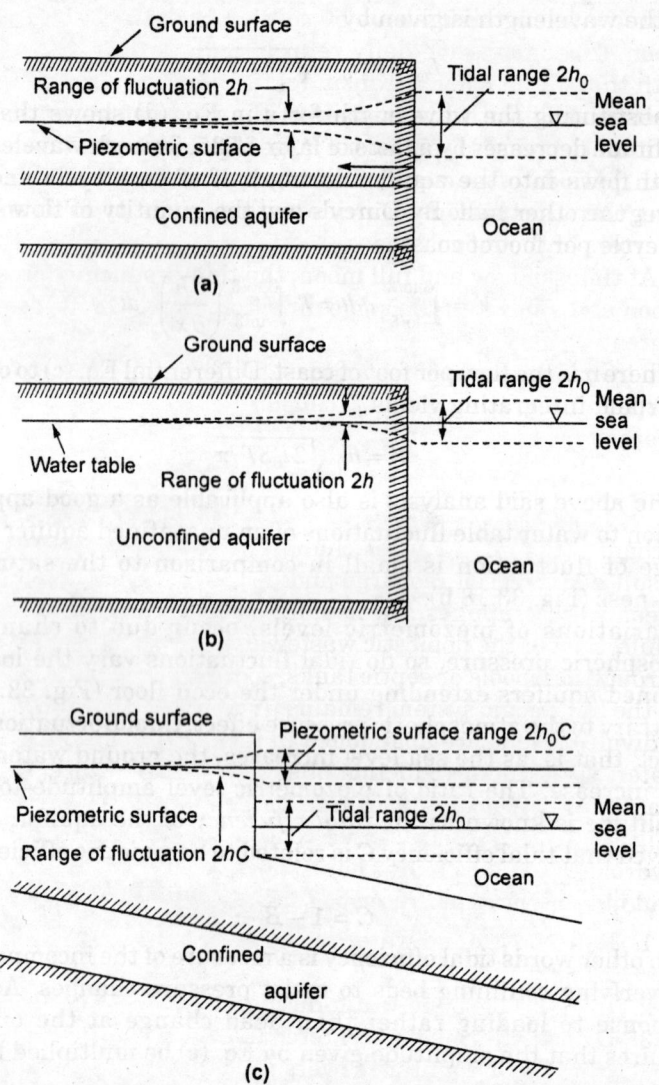

Fig. 33.18: *Ground water level fluctuations produced by ocean tides. (a) Confined aquifer, (b) unconfined aquifer, and (c) loading of a confined aquifer.*

These fluctuations result from earth tides, produced by the attraction exerted on the earth's crust by the moon and, to a lesser extent, the sun. Robinson's observations, shows (a) two daily cycles of fluctuations occur about 50 min later each day, as does the moon, (b) the average daily retardation of cycles agrees closely with that of the moon's transit, (c) the daily troughs of the water level coincide with the transits of the moon at upper and lower culmination, and (d) periods of large regular fluctuations coincide with periods of new and full moon, whereas periods of small irregular fluctuations coincide with periods of first and third quarters of the moon.

At times of new and full moon, the tide-producing forces of the moon and sun act in the same direction, then ocean tides display a greater than average range. But when the moon is in the first and third quarters, tide-producing forces of the sun and moon act perpendicular to each other, causing ocean tides of smaller than average range.

URBANIZATION

Urbanization often causes changes in ground water levels as a result of decreased recharge and increased withdrawal. In rural areas water supplies are usually obtained from shallow wells, while most of the domestic wastewater is returned to the ground through cesspools or septic tanks, so that, a quantitative balance in the hydrologic system remains. As population increases, many individual wells are abandoned in favour of deeper public wells. Later, with the introduction of sewer systems, storm water and wastewater typically discharge to a nearby surface water body (Fig. 33.19). Therefore, three conditions disrupt the subsurface hydrologic balance and produce declines in ground water levels and increase ground water pollution.

1. Reduced ground water recharge due to paved surface areas and storm sewers.
2. Increased ground water discharge by pumping wells.
3. Decreased ground water rercharge due to export of wastewater collected by sanitary sewers.

EARTHQUAKES

Earthquakes effect ground water as sudden rises or falls of water levels in wells, changes in dischagre of springs, appearance of new springs, and eruptions of water and mud out of the ground. More commonly, however, earthquake shocks produce small

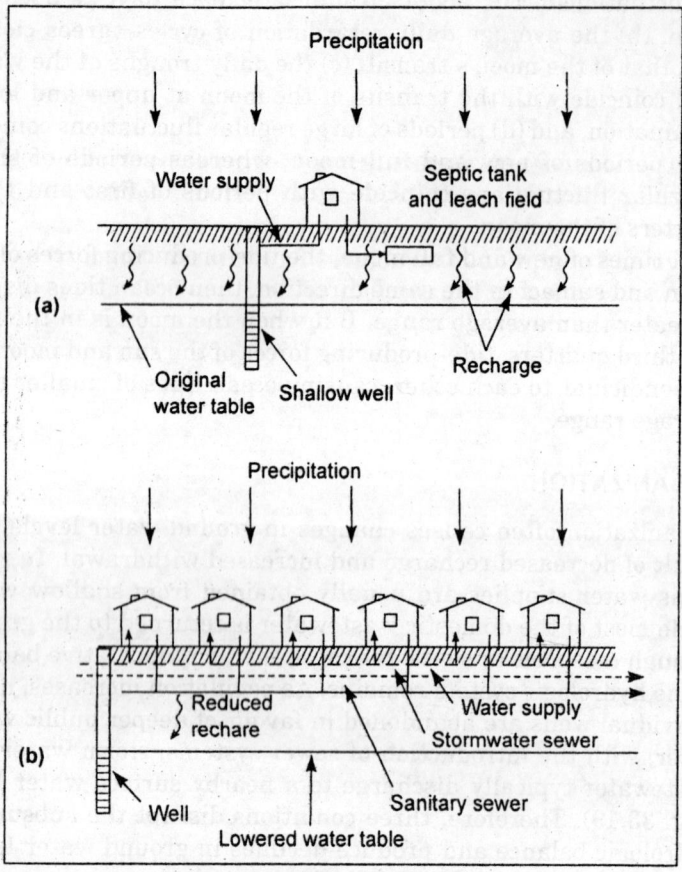

Fig. 33.19: *Schematic diagram illustrating how urbanization can cause lowering of water table elevation. (a) Rural situation and (b) urban development.*

fluctuations *(hydreseisms)* in wells penetrating confined aquifers. Quantitative effects of eartquakes on ground water is in fact little known.

But these fluctuations result from compression and expansion (dilatation) of elastic confined aquifers by the passage of earthquake (Rayleigh) waves. Looking at the converse situation, recent field studies have revealed that injection of wastewater into a deep well can trigger earthquakes. Evidence stems from injection of chemical-manufacturing waste fluids near Denver. Colorado, into a well 3671 m deep and penetrating sedimentary rocks into Pre-Cambrian crystalline rocks.

LAND SUBSIDENCE AND GROUND WATER

Ground water levels or subsurface moistrue changed conditions may be responsible for subsidence of the land surface. This can severely damage wells and can create special problems in the design and operation of structures for drainage, flood protection, and water conveyance, and four distinct phenomena have been identified for this.

Lowering of Piezometric Surface

Land subsidence has been observed to accompany extensive lowering of the piezometric surface in regions of heavy pumping from confined aquifers.

The explanation for this subsidence is based on fundamentals of soil mechanics. Consider the pressure diagram for a confined aquifer overlain by an unconfined aquifer shown in Fig. 33.20. Initially, the total (geostatic) pressure P_t at any depth (as in Fig. 33.20) is

$$P_t = P_h + P_i \qquad (a)$$

where P_h is the hydraulic pressure and P_i is the intergranular pressure. If pumping in the confined aquifer lowers the piezometric surface while the water table remains unchanged due to an impermeable clay layer separating the aquifers (Fig. 33.20 b), then Eq. (a) becomes

$$P_t = P_h + P_i \qquad (b)$$

Note that $P_h' < P_h$ and P_i, P_i' for both the confined aquifer and the clay layer. Adjustments to these new pressure distributions will take place essentially instantaneously in the permeable, coarsegrained aquifer. But in the relatively impermeable, fine-grained clay, this adjustment may take months to years, because clayey materials are highly compressible, the increased intergranular pressure are highy compressible, the increased intergranular pressure $(P_i' - P_i)$ causes the clay layer to be compacted. This reduces its porosity, while water contained in the clay pores is squeezed downward into the confined aquifer.

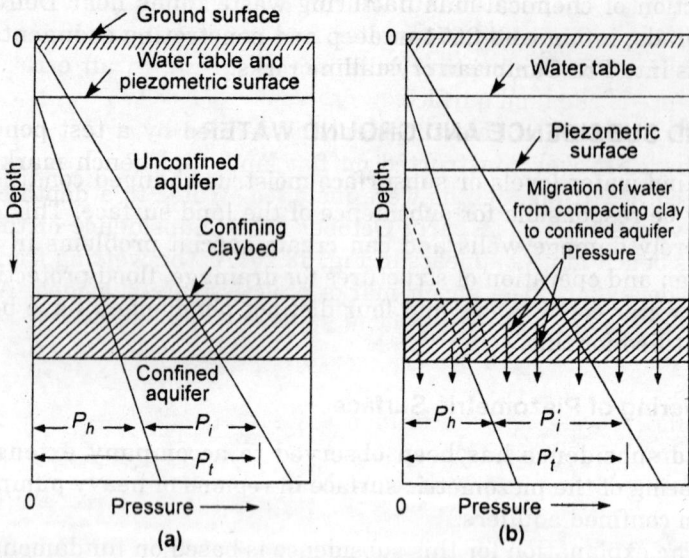

Fig. 33.20: *Graph of hydraulic and intergranular pressures as a function of depth for an unconfined aquifer overlying a confined aquifer. (a) Initial condition with water table and piezometric surface at the same elevation, and (b) subsequent conditions with piezometric surface lowered.*

The volume of water displaced from the clay equals the reduction in clay volume and also the volumetric land surface subsidence. Similarly, the reduction in thickness of the clay layer equals the vertical land subsidence. The amount of compaction is a function of the thickness and vertical permeability of the clay, of the time and magnitude of piezometric surface decline, and of the microstructure of the clay. Because sand and gravel deposits are relatively incompressible, the increased intergranular pressure has a negligible effect on the aquifer itself.

Hydrocompaction

When water is applied to certain types of soils collapse of the ground surface occur. Particularly susceptible are (a) loose, moisture-deficient alluvial deposits, including mud flows, and (b) moisture-deficient loess deposits, e.g. when soils characteristically are desiccated with a high void content and low density (1.1 to 1.4

g/cm³). Most of these soils have never been saturated since deposition, but when irrigation water, for example, is applied, their internal high void structure collapses, resulting in an erratic subsidence of the land surface.

The magnitude of this subsidence is defined by a test pond 30 m by 30 m was constructed on flat land, and bench marks anchored at various depths were installed. Water to a depth of 0.6 m was admitted in early October 1956. Subsidence of the various bench marks appears in Fig. 33.20. As the wetting front moved downward, the bench marks progressively subsided. At ground surface the change in level amounted to more than 3 m, while at the 45 m depth no effect was observed until after 16 months.

Shallow subsidence can influence irrigation, drainage, sewerage, and transportation systems. Sprinkler irrigation and pipelines for water conveyance are best suited in these terrains.

Sinkhole Formation

Catastrophic land subsidence leading to the formation of sinkholes are due to catastrophic land subsidence but it can also be associated with declines in ground water levels. Soluble rocks such as dolomite and limestone are slowly dissolved locally by ground water. Eventually, the ground surface sinks to form a cup-shaped depression. Over large areas of this type, a karstic sinkhole plain is formed with most of the drainage occurring in the subsurface.

New sinkholes often develop in regions where water tables have been lowered by pumping.

Crustal Uplift

Crustal uplift is the opposite of land subsidence which can occur over large areas subject to heavy ground water pumping. The tectonic uplift of land, involving on elastic expansion of the lithosphere, is caused by the removal of large masses of ground water.

34 Surface Investigations of Ground Water

As it is impossible to see ground water directly from the earth's surface, a variety of techniques can provide information concerning its occurrence and under certain conditions—even its quality from surface or above-surface locations. Geologic methods, involving interpretation of geologic data and field reconnaissance, represent an important first step in any ground water investigation. Remote sensing from aircraft has become an increasing valuable tool for understanding subsurface water conditions. Geophysical techniques, especially electric resistivity and seismic refraction methods, provide only indirect indications of ground water so that underground hydrologic data must be inferred from surface data. But correct interpretation requires supplemental data from subsurface investigations. There are many methods we employ to get some what positions of water under the ground so as to make it potential for the development of ground water.

METHODS OF GEOLOGIC INVESTIGATION

A geologic investigation begins with the collection, analysis, and hydrogeologic interpretation of existing topographic maps, aerial photographs, geologic maps and logs, etc. by evaluation of available hydrologic data on stream flow and springs, well yields ground water recharge, discharge, and levels and water quality. This is the first step in any investigation of subsurface water because no expensive equipment is required.

The depositional and erosional events in an area may indicate the extent and regularity of water-bearing formations. The type of rock formation suggest the magnitude of water yield. One formation may be adequate for a domestic supply but entirely unsatisfactory for an industrial or municipal supply. The nature and thickness of overlying beds, as well as the dip of water-bearing formations do estimate of drilling depths. Landforms can often reveal near surface unconsolidated formations serving as aquifers, such as glacial outwash, eskers, terraces, and sand dunes. Faults shows impermeable barriers to subsurface flow, which can be mapped from surface traces.

Remote Sensing Methods

Aerial photography of the earth at various electromagnetic wavelength ranges provide useful information regarding ground water conditions. The technology of remote sensing has developed rapidly in recent years, and has applications to water resources. Stereoscopic examination of aerial photographs has gained steadily in importance. Observable patterns, colours, and relief make it possible to distinguish differences in geology, soils, soil moisture, vegetation, and land use. Photogeology can differentiate between rock and soil types and indicate their permeability and areal distribution and areas of ground water recharge and discharge. Maps classifying an area into good, fair, and poor ground water yields can be prepared.

Aerial photographs also reveal fracture patterns in rocks, which relate to porosity, permeability, and ultimately well yield, springs and marshy areas indicate relatively shallow depths to ground water. Hydrobotanical studies of vegetation in photographs are productive. Phreatophytes, which transpire water from shallow water tables, define depths to ground water. Figure 34.1 shows vegetation on an alluvial fan. Halophytes, plants with a high tolerance for soluble salts, and white efflorescences of salt at ground surface indicate the presence of shallow brackish or saline ground water. Xerophytes, desert plants subsisting on minimal water, suggest that ground water occurrence at considerable depth.

Infrared imagery, which records differences in apparent surface temperatures, enables information on soil moisture, ground water circulation, and faults functioning as aquicludes to be obtained. Near-infrared imaging has outlined seepage patterns from canals. One of the most interesting results of infrared aerial imaging has been mapping coastal submarine springs, both hot and cold, in regions of basalt or limestone. Radar imagery can provide

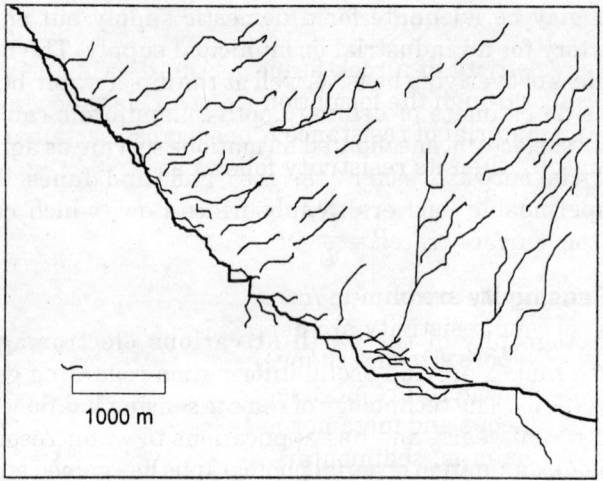

Fig. 34.1: *Tracing of an aerial photograph showing a strip of phreatophytes along the toe of an alluvial fan in a desert area. There should be a good supply of shallow ground water along the upslope portion of the strip.*

information on the presence of moisture on or at shallow depths below ground surface. Low frequency electromagnetic aerial surveys have outlined buried channels and zones of seawater intrusion.

Geophysical Exploration Method

Geophysical exploration is the scientific measurement of physical properties of the earth's crust for investigation of mineral deposits and geologic structure. The discovery of oil by geophysical methods in 1926, give economic pressures for locating petroleum and mineral deposits which stimulated the development and improvement of many geophysical methods and equipment. It's application to ground water investigations was slow because the commercial value of oil overshadows that of water. But today, many organizations concerned with ground water employ geophysical methods.

Geophysical methods detect mostly density, magnetism, elasticity, and electrical resistivity and are most commonly measured, to interpret the geologic structure, rock type and porosity, water content, and water quality.

Electric Resistivity Method

The electric resistivity of a rock formation limits the amount of current passing through the formation when an electric potential is applied. If a material of resistance R has a cross-sectional area A and a length L, then its resistivity follows as

$$P = \frac{RA}{L}$$

Units of resistivity are ohm-m²/m, or simply ohm-m.

Rock formations resistivity are depending upon the material, density, porosity, pore size and shape, water content and quality, and temperature. There are no fixed limits for resistivities of various rocks; igneous and metamorphic rocks yield values in the range 10^2 to 10^8 ohm-m; sedimentary and unconsolidated rocks, 10^0 to 10^4 ohm-m which is shown in Fig. 34.2. In porous formations, the resistivity is controlled more by water content and quality within the formation than by the rock resistivity. For aquifers

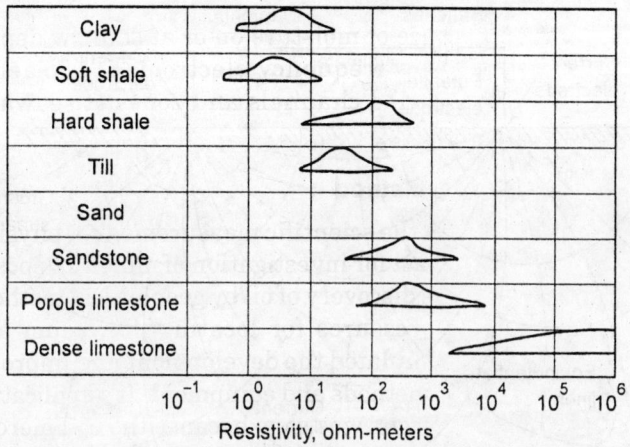

Fig. 34.2: *Representative ranges of electrical resistivity for various sediments and rocks. Values assume presence of fresh ground water; saline water will shift values at least an order of magnitude to the left.*

composed of unconsolidated materials, the resistivity decreases with the degree of saturation and the salinity of the ground water. Clay minerals conduct electric current through their matrix;

therefore, clayey formations tend to display lower resistivities than do permeable alluvial aquifers.

Clay and till when wet typically have low resistivities of 5–30 ohm-m whereas wet sand and gravel have resistivities five to ten times higher; therefore, relatively high resistivity zones are of interest as shallow aquifers.

Actual resistivities are determined from apparent resistivities, which are computed from measurements of current and potential differences between pairs of electrodes placed in the ground surface. The procedure involves measuring a potential difference between two electrodes (*P* in Fig. 34.3) resulting from an applied current through two other electrodes (*C* in Fig. 34.3) outside but in line with the potential electrodes. If the resistivity is everywhere uniform in the subsurface zone beneath the electrodes, an orthogonal network of circular arcs will be formed by the current and equipotential lines, as shown in Fig. 34.3. The measured potential difference is a weighted value over a subsurface region controlled by the shape of

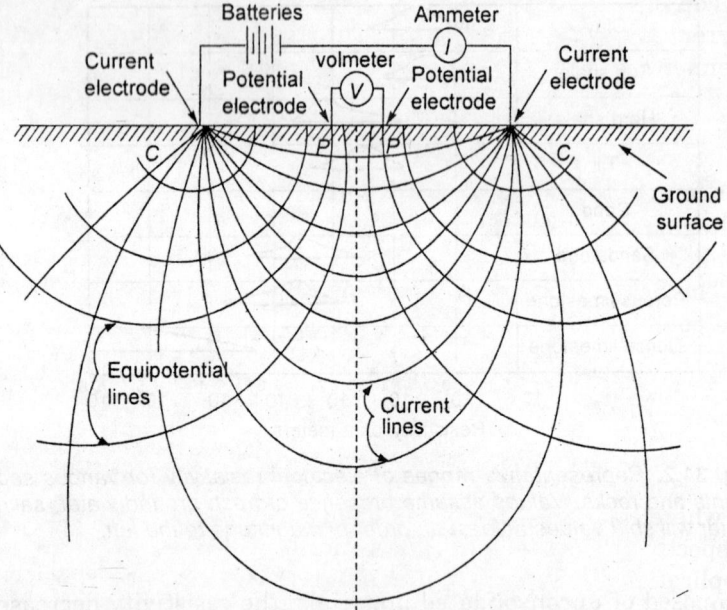

Fig. 34.3: *Electrical circuit for resistivity determination and electrical field for a homogeneous subsurface stratum.*

the network. Thus, the measured current and potential differences yield an apparent resistivity over an unspecified depth. If the spacing between electrodes is increased, a deeper penetration of the electric field occurs and a different apparent resistivity is obtained. In general, actual subsurface resistivities vary with depth; therefor, apparent resistivities generally change as electrode spacings are increased. Because changes of resistivity at great depths have only a slight effect on the apparent resistivity compared to those at shallow depths, the method is effective for determining actual resistivities below a few hundred metres depth.

Electrodes consist of metal stakes driven into the ground. For the potential electrodes, porous cups filled with a saturated solution of copper sulphate are sometimes employed to inhibit electric fields from forming around them. In practice, various standard electrode spacing arrangements are available but the Wenner and Schlumberger electrode spacing arrangements are generally adopted. Figure 34.4 (a) shows the Wenner arrangement which has the potential electrodes located at the third points between the current electrodes.

The apparent resistivity is given by the ratio of voltage to current times a spacing factor. The apparent resistivity for the Wenner arrangement as below

$$Pa = 2\pi a \frac{V}{I}$$

where a is the distance between adjacent electrodes, V is the voltage difference between the potential electrodes, and I is the applied current.

The Schlumberger arrangement, shown in Fig. 34.4 (b), has the potential electrodes close together. The apparent resistivity is given as

$$P_a = \pi \frac{(L/2)^2 - (b/2)^2}{B} \frac{V}{I}$$

where L and b are the current and potential electrode spacings, respectively (Fig. 34.4 b). Theoretically, $L > b$, but for practical application good results can be obtained if $L \geq 5b$[11].

As we plot the apparent resistivity against electrode spacing (a for Wenner, and L/2 for Schlumberger) for various spacings at

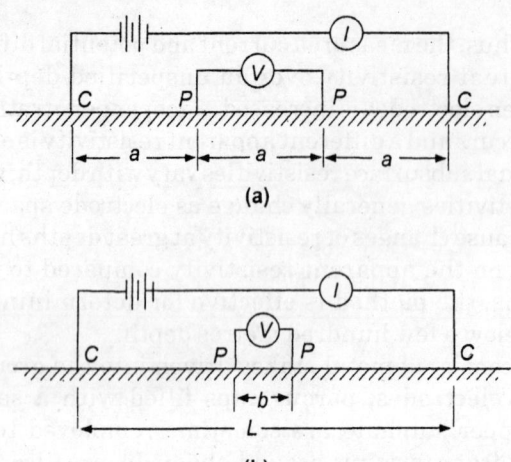

Fig. 34.4: *Common electrode arrangements for resistivity determination. (a) Wenner and (b) Schlumberger.*

one location, a smooth curve can be drawn through the points. The solution can be obtained in two parts: (a) interpretation in terms of various layers of actual (as distinguished from apparent) resistivities and their depths, (b) interpretation of the actual resistivities in terms of subsurface geologic and ground water conditions. Part (a) can be accomplished with theoretically computed resistivity-spacing curves of two-, three-, and four-layer cases for various ratios of resistivities. Part (b) depends on supplemental data. Comparing actual resistivity variations with depth to data from a nearby logged test hole enables a correlation to be established with subsurface geologic and ground water conditions. This information can then be applied for interpretation of resistivity measurements in surrounding areas.

Figure 34.5 illustrates the interpretation of a two-layer situation from measurements with the Schlumberger electrode spacing. The field curve, plotted on logarithmic transparent paper to the same scale as published master curves, is superposed on the two-layer master set. By keeping the coordinate axes paralleled, the sheet is moved until a best fit of the field and theoretical curves is obtained. The abscissa of the cross, which is the origin of the theoretical curve, equals the thickness of the first layer, while the ordinate of the cross-defines the actual resistivity P_1 of the first layer. The asymptote of the end of the curve with the largest

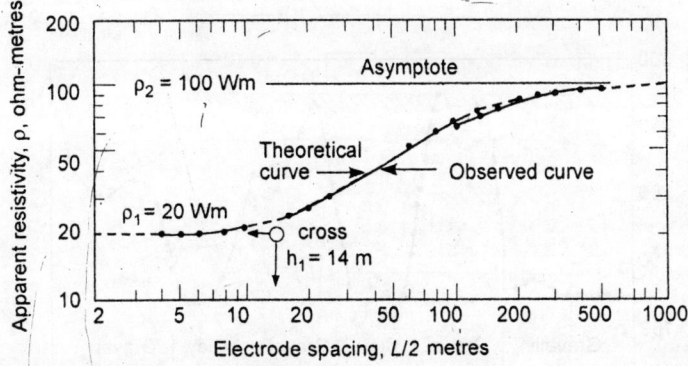

Fig. 34.5: *Interpretation of a two-layer electrical resistivity measurements from Schlumberger electrode spacings (after Zohay[59]).*

spacing defines the actual resistivity P_2 of the second layer. Physically, such a curve might represent a clay layer overlying a sandy aquifer at a depth of 14 m.

Resistivity surveys can cover vertical variations soundings at selected locations by varying electrode spacings. It is often assumed that a given electrode spacing represents the depth of resistivity measurement. Although this rule-of-thumb is untrue, the greater the current electrode separation, the greater the amount of current that penetrates a given depth.

More generally, they are conducted to obtain horizontal profiles of apparent resistivity or apparent-resistivity maps of an area by adopting a constant electrode spacing. A horizontal resistivity profile across a shallow gravel deposit together with its geologic interpretation as shown in (Fig. 34.6). Areal resistivity changes can be interpreted in terms of aquifer limits and changes in ground water quality, whereas sounding surveys may indicate aquifers, water tables, salinities, impermeable formations, and bedrock depths.

In the vicinity of electrodes the presence of lateral geologic inhomogeneities; buried pipelines, cables, and wire fences disturb the electric field. Among all surface geophysical methods, electric rsistivity has been applied most widely for ground water investigations. Its portable equipment and ease of operation facilitate rapid measurement. The method frequently aids in planning efficient and economic test drilling programs. It is especially well adapted for locating subsurface saltwater boundaries because the decrease in resistance when salt water is

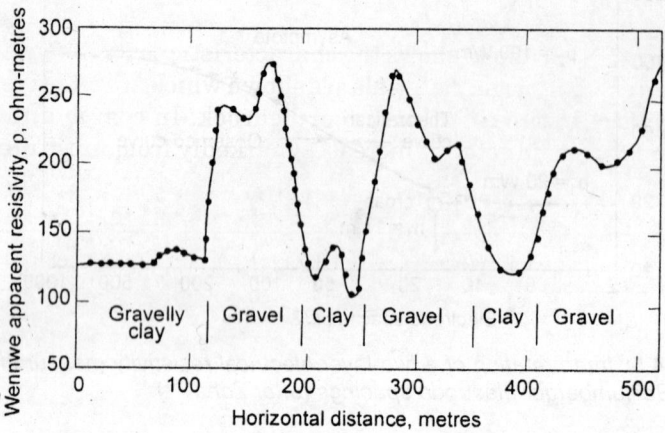

Fig. 34.6: *Horizontal profile by surface resistivity measurements over a shallow gravel deposit in California and its interpretation.*

encountered. Where subsurface conditions are relatively homogeneous, the technique can be employed to detect the water table as the top of a relatively conductive layer. The method has also been employed for delineating geothermal areas and estimating aquifer permeability.

An important new application of resistivity surveys involves defining areas and magnitudes of polluted ground water. Results correlate best with ground water samples where a highly conductive pollutant, such as soluble salt, is moving in a relatively shallow zone with uniform geologic conditions. Studies of pollution from landfills, wastewater disposal, industrial wastes, and acid mine drainage are mostly done with the help of the electric resistivity method.

Seismic Refraction Method

The creation of a small shock at the earth's surface either by the impact of a heavy instrument or by a small explosive charge and measuring the time required for the resulting sound, or shock, wave to travel known distances are generally involved by the seismic refraction method, seismic reflection methods provide information on geologic structure thoudands of metres below the surface, whereas seismic refraction methods of interest in ground water studies, go only about 100 metres deep. Seismic wave velocities are greatest in solid igneous rocks and least in unconsolidated materials.

Figure 34.7 (a and b) shows the characteristic seismic velocities for a veriety of geologic materials are shown which are employed to identify the nature of alluvium or bedrock. In coarse alluvial materials, seismic velocity increases markedly from unsaturated

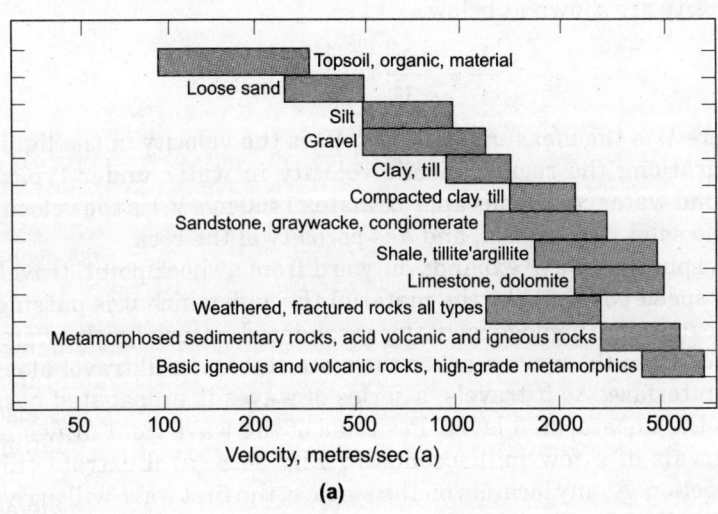

(a)

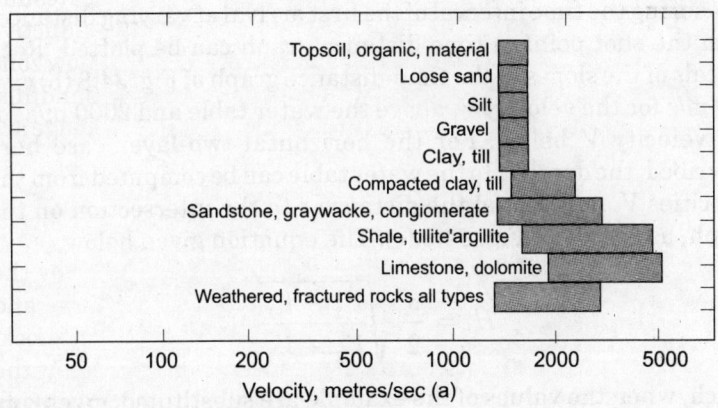

Fig. 34.7: *Seismic velocity of geologic materials. (a) Unsaturated materials, and (b) saturated materials.*

to saturated zones; consequently, the depth to water table can be mapped. Changes in seismic velocities are governed by changes in elastic properties of the formations. In sedimentary rocks, the texture and geologic history are more important than the mineral composition. Porosity tends to decrease wave velocity but increased by water content. Consolidated formations with a uniform distribution of small pores, such as a sands tone. Velocity and porosity are shown as below.

$$\frac{1}{V} = \frac{a}{V_L} + \frac{1-\alpha}{V_S}$$

where V is the measured velocity, V_L is the velocity in the liquid saturationg the rock, (Seismic velocity in water under typical ground water conditions approximates 1460 m/s V_s) is the velocity of the solid rock matrix, and a is porosity of the rock.

A spherical wave expands outward from a shock point, travels at a speed governed by the material through which it is passing. For example, a homogeneous unconsolidated material with a water table; when the wave reaches the water table it will travel along the interface. As it travels, a series of waves is propagated back into the unsaturated layer. Positions of the wave front drawn at intervals of a few milliseconds in Fig. 34.8 (a) illustrate this refraction. At any location on the surface, the first wave will arrive either directly from the shot point or from a refracted path. By measuring the time interval of the first arrival at varying distances from the shot point, a time-distance graph can be plotted. Reciprocals of the slopes in the time-distance graph of Fig. 34.8 (b) give 500 m/s for the velocity V_1 above the water table and 2000 m/s for the velocity V_2 below. For the horizontal two-layer case here described, the depth H to the water table can be computed from the velocities V_1 and V_2 and the distance s to the intersection on the graph, as shown in Fig. 34.8 (c). The equation given below,

$$H = \frac{s}{2} \sqrt{\frac{V_2 - V_1}{V_2 + V_1}}$$

which, when the values of the example are substituted, gives 8 m. Often aided by nomographs. Multilayered problems can be solved in a similar manner. Seismic refraction field investigations has been simplified with the help of compact and efficient instruments.

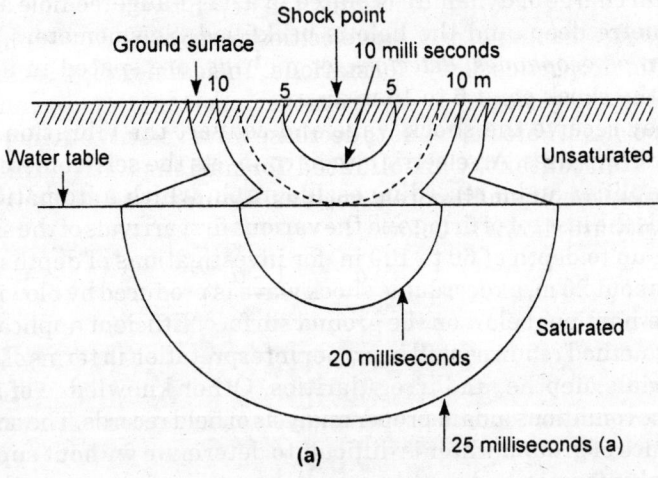

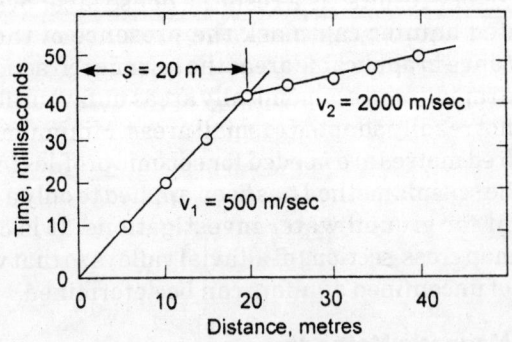

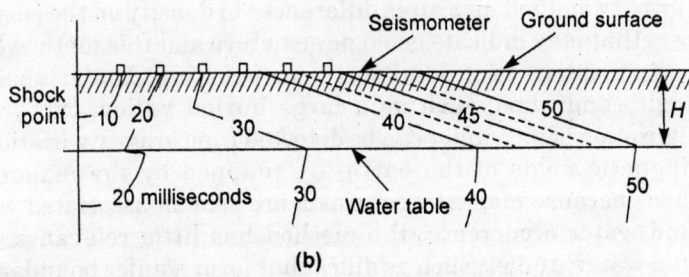

Fig. 34.8: *Seismic refraction method applied to determine depth to water table. (a) Wave front advance and (b) time–distance graph.*

A small charge of dynamite is placed in a hand-augered hole about one metre deep, and the hole is backfilled. Seismometers, also known as *geophones*, *detectors*, or *pickups*, are spaced in a line from the shock point 6 to 15 m apart.

They receive the shock wave and convert the vibration into electric impulses. An electric circuit connects the seismometers to an amplifier and a recording oscillograph, which automatically records the instant of firing and the various first arrivals of the shock wave. up to depth of 60 to 100 m for investigations of depths less than about 20 m, a recordable shock wave is produced by blowing a sledge-hammer below on the ground surface. Efficient application of the method requires skill in proper interpretation in terms of rock materials, depths, and irregularities. Other knowledge of sub-surface conditions aids in proper analysis of field records. The actual presence of ground water is difficult to determine without supple-mental information because velocities overlap in saturated and unsaturated zones. Seismic velocities must increase with depth to obtain satisfactory results; as a result, a dense layer overlying an unconsolidated aquifer can mask the presence of the aquifer. Paragraph change in applicable areas the seismic refraction method can eliminate rapidly and ecomomically areas unfavourable for test drilling. It is not readily adapted to small areas. Minimum distances of a few hundred metres are needed for seismic profiles in different directions. The scismic method has been applied to only a relatively limited extent for ground water investigations. It is commonly employed to map cross-sections of alluvial valleys so that variations in thickness of unconfined aquifers can be determined.

Gravity and Magnetic Methods

The gravity method measures differences in density on the earth's surface that may indicate geologic structure and this method has little application to ground water prospecting. Under special geologic conditions, such as a large buried valley, the gross configuration of an aquifer can be detected from gravity variations.

Magnetic fields of the earth are mapped by the magnetic methods because magnetic contrasts are seldom associated with ground water occurrence, the method has little relevance. To ground water studies, such as dikes that form aquifer boundaries or limits of a basaltic flow has been obtained with the magnetic method.

Water Witching

A forked stick used to locate water is known as *water witching* or *dowsing*. Water witches diligently practice the art wherever people can be persuaded of its potential value. Commonly, the method consists of holding a forked stick in both hands and walking over the local area until the butt end is attracted downward—ostensibly by subsurface water.

It is amazing that the idea of supernatural powers has such a continued fascination for people. A novelist Kenneth Roberts. In his novels advocating water withching (including Henry Gross and His Dowsing Rod, 1951. The Seventh Sense, 1953, and Water Unlimited, 1957) are interesting to read. The US Geological Survey advises inquirers not to employ water withches, yet the practice continues and receives frequent publicity (Fig. 34.9).

Fig. 34.9: *Water witching.*

An anthropologist has written an intriguing and well-documented analysis of water witching. This revealed an average of 181 water witches per one million population. Witching proved to be more common in rural areas and where ground water was difficult to find. The experts concluded water witching to be "magical divination," meaning an irrational system of decision-making in which the signs have no demonstrable connection to the anticipated outcome.

35 Subsurface Investigations of Ground Water

Subsurface investigations make us able to study of ground water and its condition of occurrence. We may know the location, thickness, composition, permeability, and yield of an aquifer, or location, movement and quality of ground water. Test drilling gives information on substrata in a vertical line from the surface. Geophysical logging techniques provide information on physical properties of geologic formations, water quality, well design and well development.

TEST DRILLING

To obtain assurance of underground conditions prior to well drilling, test drilling usually do of small-diameter holes. If a test hole proves fruitful, it is redrilled or reamed to a larger diameter to from a pumping well. Test holes also serve as observation wells for measuring water levels or for conducting pumping tests. In unconsolidated formations, cable tool and hydraulic rotary methods are most common. The former is slower but provides more accurate samples from the bailer; the latter is faster, but it is sometimes diffcult to determine the exact charactor of the formations because fine-grained materials are mix with the drilling fluid. For great depths and fairly uniform sands, the hydraulic rotary method is quicker and cheaper. For test holes in soft ground and shallow depths, drilling with an auger is quick and economic. Jetting has proved to be an economic method of drilling shallow, small-diameter holes for investigational purposes. The rapidity of the jetting operation combined with its lightweight portable equipment gives it important advantages, but the lack of good samples is a disadvantage.

Geologic Log

A geologic log is constructed from sampling and examination of well cuttings collected at frequent intervals during the drilling of a well or test hole. Such logs furnish a description of the geologic character and thickness of each stratum encountered as a function of depth, thereby enabling aquifers to be delineated.

Among all types of logs, the geologic log is probably the most important, but preparation of a good geologic log can be difficult. One problem is that well cuttings are small and mixed with mud. Particularly in rotary drilling, the drilling mud marks the presence of material in the silt and clay ranges.

It is good practice to store samples of well cuttings systematically. These not only permit detailed geologic logs to be prepared but also enable grain size analyses and correlations with other nearby wells to be made after drilling is finished. The technique is most practical with hydraulic rotary drilling although it is applicable to other methods as well. Because the texture of a stratum being penetrated largely governs the drilling rate, a drilling time log may be readily interpreted in terms of formation types and depths.

WATER LEVEL MEASUREMENT

In both existing and new wells water level measured data are needed to define ground water flow directions, changes in water levels overtime, and effects of pumping tests.

Lowering a steel tape into a well is a simple and accurate method for obtaining water depth. By adding chalk to the end of the tape, the length of submersion becomes apparent, thus giving the distance from the top of the well to the water surface.

For depths exceeding 50 m, an electric water-level sounder is preferable. A sounder consists of a battery, a voltmeter, a calibrated two-wire cable on a reel, and an electrode. When the electrode contacts water, the circuit is completed and the voltmeter shows a deflection. The depth is read directly from graduations along the cable.

Most widely employed technique is the air-line method. In which a small-diameter tube is placed in the annular space between the pump column and the casing. The tube is fastened to the pump column so that the two are installed simultaneously, and the tube extends below the water surface and is connected to a tire pump or small air compressor with a pressure gauge. Air is pumped into the tube until a maximum pressure is observed; this pressure, converted to depth of water, indicates the distance from the lower end of the tube to the water surface. The air-line method is less accurate than the above methods, but it is especially applicable in pumped wells where water splash and turbulence may invalidate other techniques.

Stewart developed rock technique to measure water level in deep wells. He determined empirically the time required for a common 1.55 cm glass marble or a standard BB (air rifle shot) to fall to the water surface plus the time for the sound of the splash to return to ground surface. Elapsed time measurement usually do to by stop watch. The depth to water can be read directly from Table 35.1. For more than 57 m depth the sphere reaches a constant terminal velocity; therefore, for depths greater than those listed in Table 35.1 the equation

$$d = 27.3\,t - 47.6$$

can be employed where d is the depth to water in meters and t is the time interval in seconds. This method shows accuracy up to 15 m depth. A typical recorder consists of a float and counter weight, a gear linkage that rotates a chart drum, and a recording pen driven across the chart by a clock mechanism.

Area where multiple aquifers exist with differing water levels, individual observation wells screened in only one aquifer are often drilled. Alternatively, individual small piezometer tubes extending down to the levels of the various aquifers are placed inside a large single perforated casing. The cassing is backfilled with sand and sealed by grout between adjoining aquifers.

Table 35.1 applies only to the two specified spheres; ordinary pebbles give erratic results because of their irregular shapes. Stewart also pointed out that BBs make a sharp "ping" sound, while marbles cause a short "blurred" sound. Furthermore, it is to be noted that, are deflected by spiderwebs, but marbles are not.

Table 35.1: Depth to water surface in a well as a function of the time interval of a falling sphere (after Stewart).

Time, s	Distance, m	Time, s	Distance, m	Time, s	Distance m
0.0	0.0	1.4	9.6	2.8	33.3
0.1	0.0	1.5	11.0	2.9	35.4
0.2	0.2	1.6	12.4	3.0	37.6
0.3	0.4	1.7	13.9	3.1	39.8
0.4	0.8	1.8	15.3	3.2	42.1
0.5	1.2	1.9	16.8	3.3	44.4
0.6	1.8	2.0	18.4	3.4	46.9
0.7	2.4	2.1	20.0	3.5	49.2
0.8	3.1	2.2	21.7	3.6	51.6
0.9	4.0	2.3	23.5	3.7	54.1
1.0	4.9	2.4	25.4	3.8	56.7
1.1	5.9	2.5	37.3	3.9	59.1
1.2	7.1	2.6	29.3	4.0	61.6
1.3	8.3	2.7	31.3	4.1	64.3

GEOPHYSICAL LOGGING

To interpret formation charcteristics, ground water quality, quantity and movement or in other words, physical structure of the bore hole. Geophysical logging involves lowering sensing devices in a bore hole and recording a physical parameter. Table 35.2 shows the types of information that can be obtained from various logging techniques.

Table 35.2: Summary of logging applications to ground water hydrology (after Keys and MacCrary)

Required information	Possible logging techniques
Lithology and stratigraphic correlation of aquifers and associated rocks	Resistivity, sonic, or caliper logs made in open holes; radiation logs made in open or cased holes.
Total porosity or bulk density	Calibrated sonic logs in open holes; calibrated neutron or gamma-gamma logs in open or cased holes.

Table 35.2: *(Contd.)*

Required information	Possible logging techniques
Effective porosity or true resistivity	Calibrated long-normal resistivity logs
Clay or shale content	Natural gamma logs
Permeability	Under some conditions long- normal resistivity logs.
Secondary permeability- fractures, solution openings	Caliper, sonic, or television logs
Specific yield of unconfined aquifers	Calibrated neutron logs
Grain size	Possible relation to formation factor derived from resistivity logs
Location of water level or saturated zones	Resistivity, temperature, or fluid conductivity logs; neutron or gamma-gamma logs in open or cased holes
Moisture content	Calibrated neutron logs
Infiltration	Time-interval neutron logs
Dispersion, dilution, and movement of waste	Fluid conductivity or temperature logs; natural gamma logs for some radioactive wastes
Source and movement of water in a well	Fluid velocity or temperature logs
Chemical and physical characteristics of water, including salinity, temperature, density and viscosity	Calibrated fluid conductivity or temperature logs; resistivity logs
Construction of existing wells, diameter and position of casing, Perforations, screens	Gamma-gamma, caliper, casing, or television logs
Guide to screen setting	All logs providing data on the litho- logy, water-bearing characteristics, and correlation and thickness of aquifers
Cementing	Caliper, temperature, or gamma- gamma logs; acoustic logs for ce- ment bond
Casing corrosion	Under some conditions caliper, casing, or television logs
Casing leaks and/or plugged screen	Fluid velocity logs

To correlate one well to another, geophysical logs furnish continuous records of subsurface conditions which serve as valuable supplements to geologic logs and data from geophysical logs can be digitized, stored on magnetic tape, or transmitted by radio or telephone for interpretation. Graphic displays of log data permit rapid visual interpretations and comparisons in the field so that it helps in completion and testing of wells can be made immediately (Fig. 35.1).

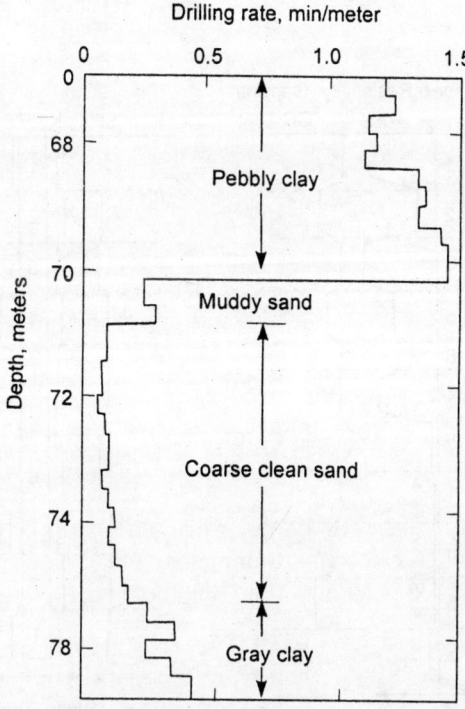

Fig. 35.1: *Drilling-time log and strata penetrated.*

Most water wells are shallow, small-diameter holes for domestic water supply; logging costs would be relatively large and usually unnecessary. But for deeper and more expensive wells, such as for municipal, irrigation, or injection purposes, logging can be economically justified in terms of improved well construction and performance.

As logging techniques become more sophisticated, the data they produce become more complex. The interpretation of many logs is more of an art than a science; log responses are governed by numerous environmental factors, making quantitative analysis difficult. In general, best results are obtained with experience and with supplemental hydrogeologic information.

Figure 35.2 is a schematic diagram showing several of the logs and their typical relative responses in various unconsolidated and consolidated geologic formations.

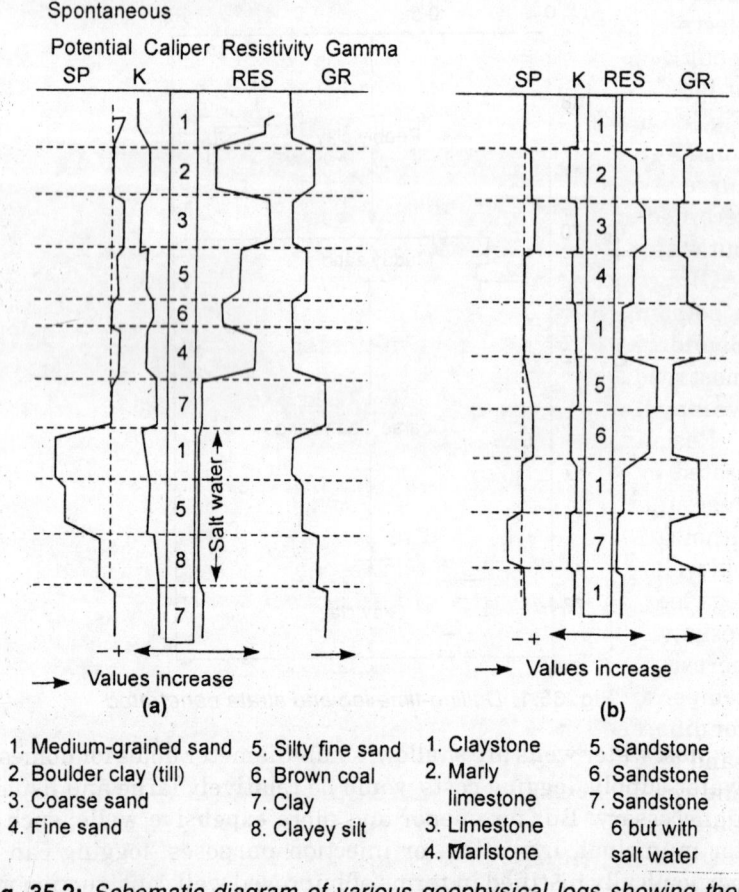

Fig. 35.2: *Schematic diagram of various geophysical logs showing their relative responses in (a) unconsolidated rocks and (b) consolidated rocks.*

Resistivity Logging

To measure electric resistivities of the surrounding medium and to obtain a trace of their variation with depth. Current and potential electrode are lowered in an uncasd well. The result is a resistivity (or electric) log. Such a log is affected by fluid within a well, by well diameter, by the character of surrounding strata, and by ground water.

The multielectrode method is most commonly employed for measuring underground resistivities because it minimizes effects of the drilling fluid and well diameter and also makes possible a direct comparison of several recorded resistivity curves. Four electrodes, two for emitting current and two for potential measurement, constitute the system. Recorded curves are termed normal or lateral, depending on the electrode arrangement, as shown in Fig. 35.3. In the normal arrangement the effective spacing is considered to be the distance AM (Fig. 35.3 a) and the recorded curve is designated AM. Sometimes a long normal curve (AM') is recorded based on the same electrode arrangement as the normal but with a larger AM distance (Fig. 35.3 b). The spacing for lateral (AO) curves is taken as the distance AO, measured between A and a point midway between the electrodes M and N (Fig. 35.3 c). Boundaries of formations having different resistivities are located most readily with a short electrode spacing, but with long spacings, we may obtain information on fluids in thick permeable formation.

Resistivity curves indicate the lithology of rock strata penetrated by the well. Fresh and salt waters distinguished in the surrounding material by resistivity curves which indicate the lithology of rock strata penetarated the surrounding material. In old wells exact locations of casings can be determined. Resistivity logs may be used to determine specific resistivities of strata. As resistivity of an unconsolidated aquifer is controlled primarily by porosity, packing, water resistivity, degree of saturation, and temperature. Although specific resistivity values cannot be stated for different aquifers, on a relative basis shale, clay, and saltwater sand give low values, fresh-water sand moderate to high values, and cemented sandstone and nonporous limestone high values. Casings and metallic objects will indicate very low resistivities, correlation of rock samples, taken from wells during drilling, with resistivity curves furnishes a sound basis for interpretation of curves measured in nearby wells without available samples.

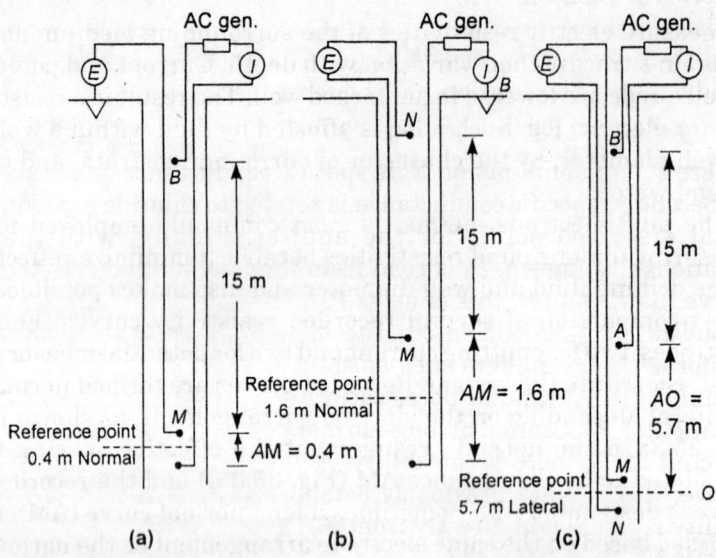

Fig. 35.3: *Typical electrode arrangements and standardized distances for resistivity logs. (a) Short normal, (b) long normal, and (c) lateral.*

Resistivity of ground water depends on ionic concentration and ionic mobility of the salt solution. This mobility is related to the molecular weight and electrical charge, so that differences exist for various compounds. As a greater ionic mobility associated with a decrease in viscosity happens due to increase in temperature of ground water. Hence, and inverse relation exists between resistivity and temperature. Resistivity at the measurement temperature when multiplied by the correction factor for that temperature yields the resistivity at the standard temperature of 25° C.

Most common uses of an electric log is to determine the proper place to set well screens. A log provides a basis for selecting proper lengths of screens and for setting them opposite the best formations. The applicability of resistivity logs to the estimation of ground water quality made by the investigations of Louisiana aquifers by Jones and Buford and later by Turcan. A field-formation factor F for an aquifer is determined as below

$$F = \frac{P_o}{P_w}$$

where P_o is the resistivity of the saturated aquifer and P_w is the resistivity of the ground water in the aquifer. It should be noted that the resistivity of ground water is the reciprocal of its specific conductance. The relation has the form

$$P_w = 10^4/E_c$$

where P_w is in ohm-m and E_c is specific conductance in uS/cm.

Secondly, specific conductance is related to chloride content or total dissolved solids for the aquifer. Finally, with these relationships known, P_o is read from the long-normal resistivity curve in an aquifer; this enables P_w and then the salinity of the groudwater to be calculated. The method yields best results in uniform clastic aquifers such as sand and sandstone, consisting mainly of intergranular pores saturated with water.

Another application of long-normal curves has been suggested by croft to estimate permeability. A value of F is determined as above; then from a previously established relationship, permeability is calculated directly from F. The method is applicable to clastic rock formations.

Resistivity logs can also aid in identifying wells that intersect both fresh and saline zones. Circulation within such a well under non-pumping conditions depends on the relative hydrostatic heads, water densities, aquifer locations and thicknesses, and the physical structure and condition of the well.

SPONTANEOUS POTENTIAL LOGGING

The spontaneous potential or, self potential (SP) method measures natural electrical potentials found within the earth. Usually measured in millivolts are obtained from a recording potentiometer connected to two like electrodes. One electrode is lowered in an uncased well and the other is connected to the ground surface, as illustrated by electrodes M and N in Fig. 35.3 (a). The potentials are primarily produced by electrochemical cells formed by the electrical condutivity differences of drilling mud and ground water where boundaries of permeable zones intersect a borehole. Potential logs indicate permeable zones but not in absolute terms; they can also aid in determining casing lengths and in estimating total dissolved solids in ground water. Where no sharp contrasts occur in permeable zones, as often happens in shallow alluvial formations, potential logs lack relief and contribute little. In urban and industrial areas, spurious earth currents may occur, such as from electric railroads, which interfere with potential logging.

By convention potential logs are read in terms of positive and negative deflections from and arbitrary baseline, usually associated with an impermeable formation of considerable thickness. The sign of the potential depends on the ratio of the salinity (or resistivity) of the drilling mud to the formation water.

Spontaneous potentials resulting from electrochemical potentials can be expressed by

$$SP = - (64.3 + 0.239\ T) \log \frac{P_f}{P_w}$$

where P_f is the drilling fluid resistivity in ohm-m, P_w is the ground water resistivity in ohm-m, and T is the borehole temperature in °C. Therefore, for measured SP, P_f and T values, the resistivity and hence salinity of ground water can be determined. It should be noted, however, that the formula applies only where the ground water is very saline.

RADIATION LOGGING

Radiation logging, also known as *nuclear* or *radioactive logging*, involves the measurement of fundamental particles emitted from unstable radioactive isotopes. Logs having application to ground water are natural-gamma, gamma-gamma, and neutron, and generally recorded in either cased or open holes which are filled with any fluid.

Natural-Gamma Logging

The radiation originates from unstable isotopes of potassium, uranium, and thorium. The natural-gamma activity of clayey formations is significantly higher than that of quartz sands and carbonate rocks. The most important application to ground water hydrology is identification of lithology, particularly clayey or shale-bearing sediments, which posses the highest gamma intensity.

Figure 35.4 shows the natural-gamma log of a test hole in unconsolidated sediments together with its geologic interpretation.

Primary applicatio̅ns of gamma-gamma logs are for identification of lithology and measurement of bulk density and porosity of rocks. The porosity α can be determined by

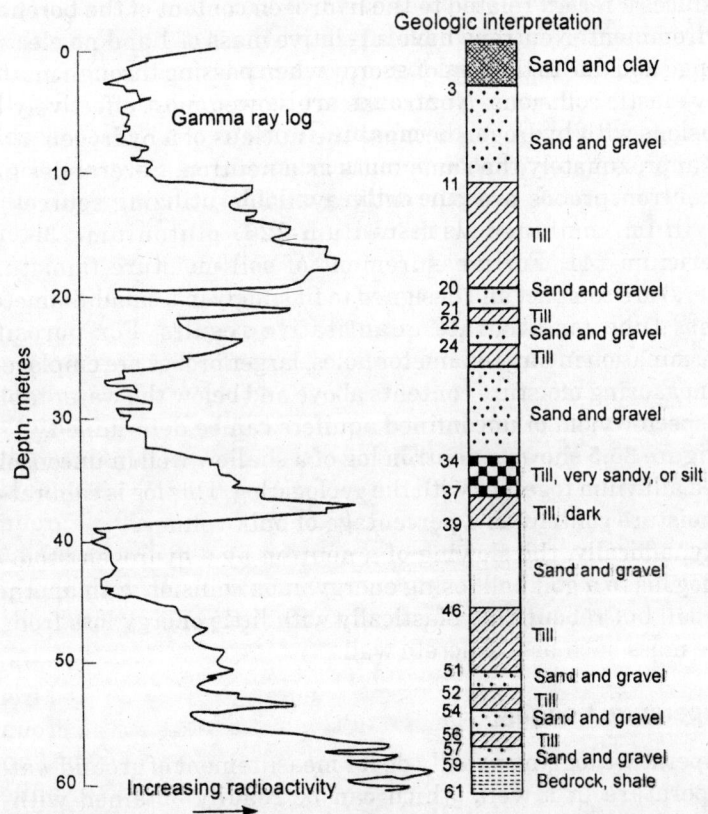

Fig. 35.4: *Natural-gamma log of a test hole in Moraine city, Ohio, together with its geologic interpretation.*

$$\alpha = \frac{P_G - P_B}{P_G - P_f}$$

where P_G is grain density (of cuttings or cores) P_B is bulk density which is measured from a calibrated log and P_f is the fluid density.

Neutron Logging

Neutron logs measure moisture content above the water table and porosity below the water table and this logging is accomplished by a neutron source and detector arranged in a single probe, which produces a record related to the hydrogen content of the borehole enviromment. Neutrons have a relative mass of 1 and no electric change and due to the loss of energy when passing through matter is by elastic collisions. Neutrouns are slowed most effectively by collisions with hydrogen because the nucleus of a hydrogen atom has approximately the same mass as a neutron. Several designs of neutron probes are currently available utilizing sources of beryllium combined with radium-226, plutonium-239, or americium-241. For measurement of soil moisture (moisture meters) are compact and designed to fit snugly in a small-diameter access tube for accurate quantitative results. For porosity determination in large-diameter holes, larger probes are employed. By measuring moisture contents above and below the water table, the specific yield of unconfined aquifers can be determined.

Figure 35.5 shows a neutron log of a shallow well in unconsolidated alluvium together with the geologic log. This log is calibrated in moisture content as a percentage of bulk volume.

Dynamically, the slowing of a neutron by a hydrogen atom is analogous to a golf ball losing energy upon collision with another golf ball but rebounding elastically with little energy loss from a large mass such as a concrete wall.

Temperature Logging

Temperature logging usually do for measurement of ground water temperature in a well, which can be readily obtained with a recording resistance thermometer for analyzing subsurface conditions. Temperatures increase with depth in accordance with the geothermal gradient, as roughly $30°C$ for each 100 m in depth. Departures from this normal gradient provide information on circulation or geologic conditions in the well. Cold temperatures may indicate the presence of gas or, in deep wells, may suggest recharge from ground. Temperatures may indicate waters from different aquifers intersected by a well.

Caliper Logging

A caliper log provides a record of the average diameter of a borehole. These logs aid in the identification of lithology and stratigraphic correlation, in the location of fractures and other rock openings, and in correcting other logs for hole-diameter

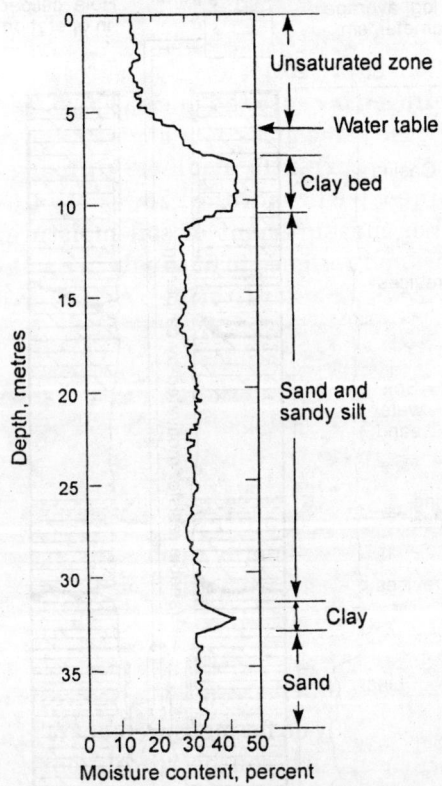

Fig. 35.5: *Neutron log of a shallow well in unconsolidated alluvium near Garden city, Kansas, together with its geologic interpretation.*

effects. During well construction caliper logs indicate the size of casing that can be fitted into the hole and enable the annular volume for garvel packing to be calculated. Other applications include measuring casing diameters in old wells and locating swelling and caving zones as shown in Fig (35.6).

Fluid-Conductivity Logging

Fluid resistivity is generally measured in ohm-m; its reciprocal, conductiviy, is measured in us/cm as in equation

$$P_w = 10^4/E_C$$

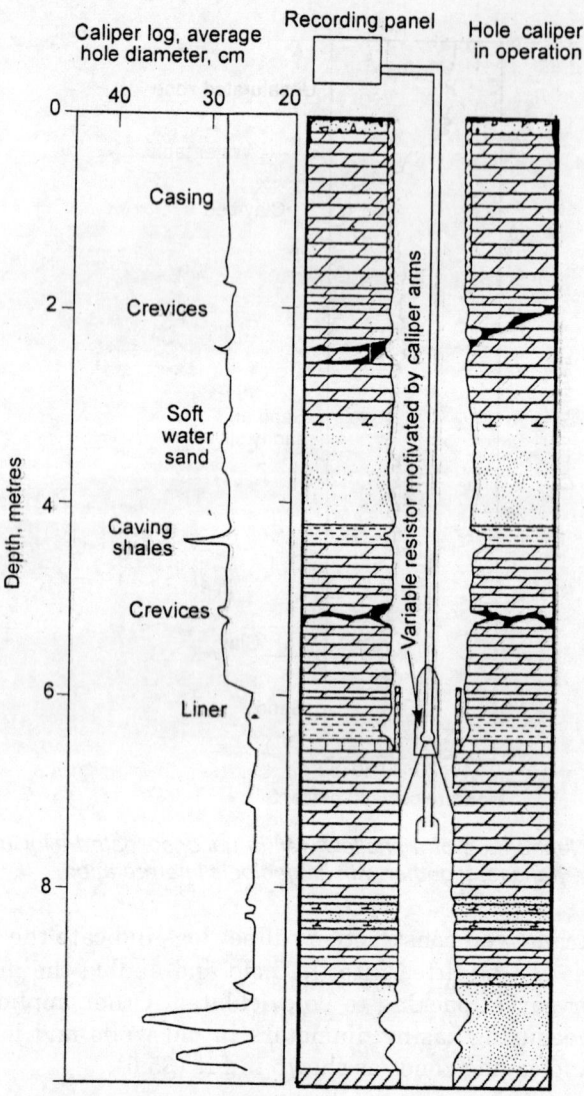

Fig. 35.6: *Hole caliper and corresponding caliper log.*

Fluid-conductivity logs enable saline water zones to be located, furnish information on fluid within a well and provide a means to extrapolate water.

FLUID-VELOCITY LOGGING

Fluid-velocity logging reveals strata contributing water to a well, flow from one stratum to another within a well, hydraulic differences between aquifers intersected by a well, and casing leaks.

OTHER IMPORTANT LOGGING TECHNIQUES

Television Logging

Specially designed wide-angle cameras, typically less than 7 cm in diameter, are equipped with lights and provide continuous visual inspection of a borehole; with videotape a record of the interior can be preserved. This is mostly applicable in locating changes in geologic strata, pinpointing large pore spaces, inspecting the condition of well casing and screen, checking for debris in wells, locating zones of sand entrance, and searching for lost drilling tools.

Acoustic Logging

Acoustic, or sonic, logging measures the sound velocity in rock which is governed by the velocity of the rock matrix and the fluid filling the pore space, therefore, the greater the porosity, the closer the measured sound velocity approaches that of the fluid. This logging widely applicable in determining the depth and thickness of porous zones estimating porosity, identifying fracture zones, and the bonding of cement between the casing and the formation are also determined by acoustic logging.

Casing Logging

A casing-collar locator is a useful instrument which consists of a magnet wrapped with a coil of wire; voltage fluctuations caused by changes in the mass of metal cutting the lines of flux from the magnet are recorded to form the log. Casing logging is useful in locating casing collars, perforation, casing and screens.

OTHER SUBSURFACE METHODS

Other subsurface methods also yield important information about hydrogeologic conditions. These are listed below.

1. Ground water level fluctuation measurements for aquifer characteristics.
2. Tracer tests for ground water flow.
3. Ground water samples for water quality determination.
4. Ground water level measurements for flow directions and aquifer conditions.
5. Pumping tests of wells for aquifer characteristics.

36 | Intrusion of Saline Water in Aquifers

The most common pollutant of fresh ground water is saline water. Intrusion of saline water occurs where saline water displaces or mixes with fresh-water in an aquifer. Its occurence takes place in deep aquifers with the upward advance of saline waters of geologic origin, in shallow aquifers from surface waste discharges and in coastal aquifers from seawater invasion. Management techniques that enable development of fresh-water and at the same time control of saline intrusion are available, Which we get through field study and theoretically also.

OCCURRENCE OF SALINE WATER INTRUSION

Generelly, human activities are main responsible for saline water intrusion into fresh ground water formations. It should be noted that saline ground water can represent a valuable resource, particularly in arid inland regions. Potential uses include industrial processes such as cooling irrigation, where the mineral content is moderate, and desalination for local domestic purposes.

As most large sources of fresh ground water are in close proximity to sea. Shallow fresh-water overlies saline water because the flushing action during recent time removes salts from ancient marine deposits. But deep ground water movement is much less so that displacement of saline water is slower.

Furthermore, at depths of several thousand metres, brines are normally encountered. Survey shows that approximately two-thirds of the United States is underlain by aquifers known to produce 1000 mg/1 saline water. There are the several sources of saline water in aquifer.

1. Human saline wastes.
2. Encroachment of seawater in coastal areas.
3. Seawater that entered aquifers during past geologic time.
4. Water concentrated by evaporation in tidal lagoons, playas, or other enclosed areas.
5. Salt in salt domes, thin beds, or disseminated in geologic formations.
6. Return flows to streams from irrigated lands.

There are three categories of mechanisms which are responsible for saline water intrusion: (a) the reduction or reversal of ground water gradients, which permits denser saline water to displace fresh-water, which commonly occurs in coastal aquifers in hydraulic continuity with the sea when pumping of wells disturbs the natural hydrodynamic balance, (b) the destruction of natural barriers that separate fresh and saline waters, e.g. the construction of a coastal drainage canal that enables tidal water to advance inland and to percolate into a fresh-water aquifer, (c) there is sub-surface disposal of waste saline water, such as into disposal wells, landfills, or other wastes.

The occurrence of saline water intrusion is extensive and represents a special category of ground water pollution, reveals that the problem exists in localities of most parts of the united States. Seawater intrusion along coasts has received the most attention. Interantionally, the problem has received attention in populated coastal areas in England, Germany, Netherlands. Israel, Japan and India also.

GHYBEN-HERZBERG RELATION BETWEEN FRESH AND SALINE WATERS

Two investigators, viz. W.B. Ghyben and A. Herzberg, working independently along the European coast, found that salt water occured underground, not at sea level but at a depth below sea level of about 40 times the height of the fresh-water above sea level. This distribution was atributed to a hydrostatic equilibrium existing between the two fluids of different densities. The equation referred to as the Ghyben-Herzberg relation.

Almost unnoticed in the hydrologic literature is the much earlier contribution of Joseph DüCommun, a French teacher at West Point Military Academy, he first stated clearly and correctly the fresh-salt water balance existing in coastal aquifers.

The hydrostatic balance between fresh and saline waters can be illustrated by the U-tube shown in Fig. 36.1. Pressures on each side of the tube must be equal, therefore

$$P_s g h_s = p_f g \, (z + h_f)$$

where P_s is the density of the saline water, P_f is the density of the fresh-water, g is the acceleration of gravity, and z and h_f are as shown in Fig. 36.1. Solving for z yields

$$z = \frac{P_f}{P_s - P_f} \, h_f \qquad \text{(b)}$$

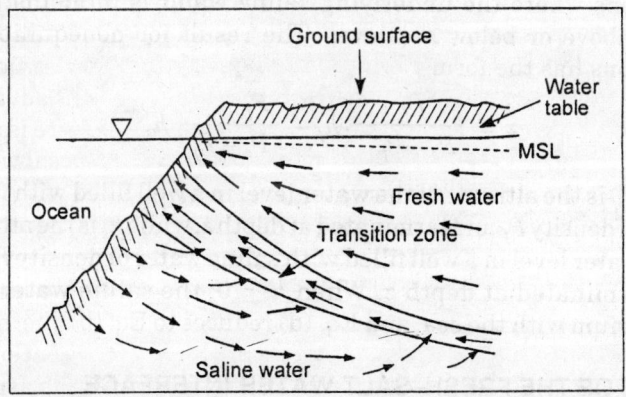

Fig. 36.1: *Vertical cross-section flow patterns of fresh and saline water in an unconfined coastal aquifer.*

Above is the Ghyben-Herzberg relation. For typical seawater conditions, let $P_s = 1.025$ g/cm^3 and $P_f = 1.000$ g/cm^3, so that

$$z = 40 \, h_f \qquad \text{(c)}$$

U-tube to a coastal situation, h_f becomes the elevation of the water table above sea level and z is the depth to the fresh-saline interface below sea level. This is a hydrodynamic rather than a hydrostatic balance because fresh-water is flowing towards the sea. From density considerations alone, without flow, a horizontal interface would develop with fresh-water everywhere floating above saline water. It can be shown that where the flow is nearly horizontal, the Ghyben-Herzberg relation gives satisfactory results. Only near the shoreline, where vertical flow components become pronounced do not give satisfactory results but errors occur at the position of interface.

The above derivation for confined aquifers can also be applied by replacing the water table by the piezometric surface. It is important to note from the Ghyben-Herzberg relation that fresh-salt water equilibrium requires that the water table, or piezometric surface, (a) lie above sea level, and (b) slope downward toward the ocean. Without these conditions, seawater will advance directly inland.

Starting from the work of Hubbert, the Ghyben-Herzberg relation has been generalised by Lusczynski and others for situations where the underlying saline water is in motion with heads above or below sea level. The result for nonequilibrium conditions has the form

$$z = \frac{P_f}{P_s - P_f} \, h_f - \frac{P_f}{P_s - P_f} \, h_s \qquad \text{(d)}$$

where h_f is the altitude of the water level in a well filled with fresh-water of density P_f and terminated at depth z, while h_s is the attitude of the water level in a well filled with saline water of density P_s and also terminated at depth z. When $h_s = 0$, the saline water is in equilibrium with the sea, and Eq. (d) reduces to Eq. (b).

SHAPE OF THE FRESH–SALT WATER INTERFACE

More exact solutions for the shape of the fresh–salt water interface have been developed from potential flow theory. The result by glover has the form

$$z^2 = \frac{2 \, pqx}{\Delta pk} + \left(\frac{pq}{\Delta pk} \right)^2 \qquad \text{(e)}$$

where z and x $\Delta p = p_s - p_f$, k is hydraulic conductivity of the aquifer, and q is the fresh-water flow per unit length of shoreline. The corresponding shape for the water table is given by

$$h_f \left(\frac{2\Delta pqx}{p + \Delta pk} \right)^{1/2} \qquad \text{(f)}$$

The width x_0 of the submarine zone through which fresh-water discharges into the sea can be obtained for $z = 0$, yielding

$$x_0 \left(\frac{pq}{2\Delta\, pk} \right) \qquad\qquad (g)$$

The depth of the interface beneath the shoreline z_0 occurs, where $x = 0$ (see Fig. 36.2) so that

$$z_0 \left(\frac{pq}{\Delta pk} \right) \qquad\qquad (h)$$

STRUCTUE OF THE FRESH–SALT WATER INTERFACE

Interfacial boundary described above between fresh and saline water does not occur sharply under field conditions. This zone develops from dispersion by flow of the fresh-water plus

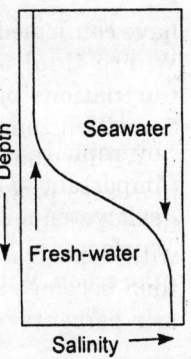

Fig. 36.2: *Increase in salinity with depth through the transition zone, linear scale.*

unsteady displacements of the interface by external influences such as tides, recharge, and pumping of wells. In general, greatest thicknessers of transition zones are found in highly permeable coastal aquifers subject to heavy pumping. Observed thicknesses very from less than 1 m to more than 100 m.

The transition zone and its seaward flow is the transport of saline water to the sea. This water originates from the underlying saline water; hence, from continuity considerations, there must exist a small landward flow in the saline water region. Figure 36.3 schematically illustrates the flow patterns in the three subsurface zones. Field measurements and experimental studies

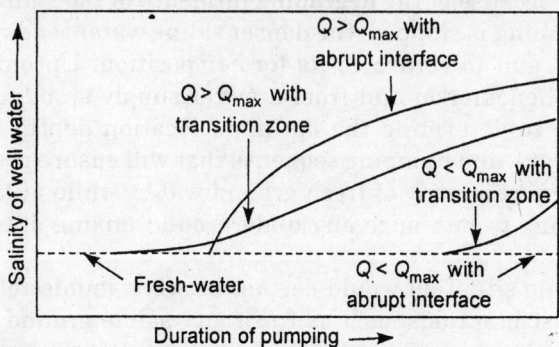

Fig. 36.3: *Well water salinity curves for upcoming of an abrupt interface and a transition zone.*

have confirmed the landward movement of the saline water body. Where tidal action is the predominant mixing mechanism, fluctuations of ground water, and hence the thickness of the transition zone become greatest near the shoreling.

Within the transition zone the salinity of the ground water increases progressively with depth from that of the fresh-water to that of the saline water. Typically, the distribution of salinity with depth varies as an error function, as shown in Fig. 36.2. It then becomes advantageous to calculate the relative salinity S_R as a percentage by

$$S_R = 100 \left(\frac{C - C_f}{C_s - C_f} \right) \tag{i}$$

where C is the salinity (salinity can be measured as total dissolved solids, chloride, or electrical conductivity) at a particular depth within the transition zone, and C_f and C_s are the salinities of the fresh and saline waters respectively.

SALINE WATER UPCOMING

When a layer of saline water below and aquifer occur and is pumped by a well penetrating only the upper fresh-water potion of the aquifer, a local rise of the interface below the well occurs. This phenomenon, known as *upcoming*. The interface is horizontal at the start of pumping when $t = t_0$. With continued pumping the interface rises to successively higher levels until eventually it can reach the well. This generally necessitates the well having to be shutdown because of the degrading influence of the saline water. When pumping is stopped, the denser saline water tends to settle downward and to return to its former position. Upcoming is a complex phenomenon and from a water-supply standpoint it is important to determine the optimum location depth, spacing, pumping rate, and pumping sequence that will ensure production of the largest quantity of fresh ground water, while at the same time striving to minimize any underground mixing of the fresh saline water.

Upcoming situation would pertain between immiscible fluids, but for miscible fluids such as fresh and saline ground water a mixing, or transition, zone having a finite thickness occurs. Although an abrupt interface neglects the physical reality of a transition zone found in ground water, the assumption has the advantage of simplicity. Furthermore, an interface can be considered as an approximation to the position of the 50 percent relative salinity in a transition zone.

An approximate analytic solution for the upcoming directly beneath a well, based on the Dupuit assumptions and the Ghyben-Herzberg relation, is given by

$$Z = \frac{Q}{2\pi dk \; (\Delta p/p_f)} \qquad (j)$$

where $\Delta p = p_s - p_f$, k is the hydraulic conductivity, and all other quantities are defined. This equation indicates an ultimate rise of the interface to a new equilibrium position that is directly proportional to the pumping rate Q.

Hydraulic model experiments have revealed that the relation in Eq. (j) holds only if the rise is limited. If the upcoming exceeds a certain critical rise, it accelerates upward to the well. The critical rise has been estimated to approximate $z/d = 0.3$ to 0.5. Thus, adopting an upper limit of $z/d = 0.5$, it follows that the maximum permissible pumping rate without salt entering the well is

$$Q_{\max} \leq \pi d^2 \, K \, (\Delta p/p_f) \qquad (k)$$

For anisotropic aquifers where the vertical permeability is less than the horizontal, a maximum well discharge larger than that for the isotropic case is possible. A transition zone with a finite thickness of brackish water occurs above the body of undiluted saline water. Upward movement of the almost fresh-water occurs readily along with the adjoining fresh-water; consequently, even with a relatively low pumping rate, no limiting critcal rise exists above which saline water will not rise. It follows that with any rate of continuous pumping, some saline water must sooner or later reach a well; equation (j) supports the previous statements. In a transition zone, the salinity changes gradually; hence, Δp in an incremental width at the top of the zone approaches zero. In equation (k) as Δp approaches zero so also must $Q_{\max}$ approach zero.

A comparison of the arrivals of salinity at a pumping well for an abrupt interface and for a transition zone is shown qualitatively in Fig. 36.3. With an abrupt interface, assuming $Q > Q_{\max}$ the salinity appears later and increases more rapidly than with a transition zone. For $Q < Q_{max}$ there will be no salinity reaching the well in the abrupt case; however, a gradual invasion of saline water will occur from a rising transition zone. The ultimate well-water salinity with upcoming approaches an intermediate value between the extremes of the fresh and saline waters; empirical data indicate this lies in the range of 5 to 8 percent of the salt concentration in the saline water.

Analysis of the above shows upcoming can be minimized by the proper design and operation of wells and galleries. For given aquifer conditions wells should be separated as far as possible vertically from the saline zone and pumped at a low uniform rate. Well and gallery designs for regions where a thin layer of fresh-water overlies saline water are discussed in the section on oceanic islands. Field tests have shown that *scavenger wells,* which pump saline water from below the fresh-water can also successfully counteract upcoming.

OCEANIC ISLANDS SHOW FRESH–SALT WATER RELATIONS

Permeable nature of oceanic islands consisting of sand, lava, coral, or limestone, so that seawater is in contact with ground water on all sides, because fresh ground water originates entirely from rainfall and only a limited quantity is available, a fresh-water lens as shown in Fig. 36.4 is formed by the radial movement of the fresh-water, toward the coast. This lens floats on the underlying salt water, its thickness decreases from the centre towards the coast.

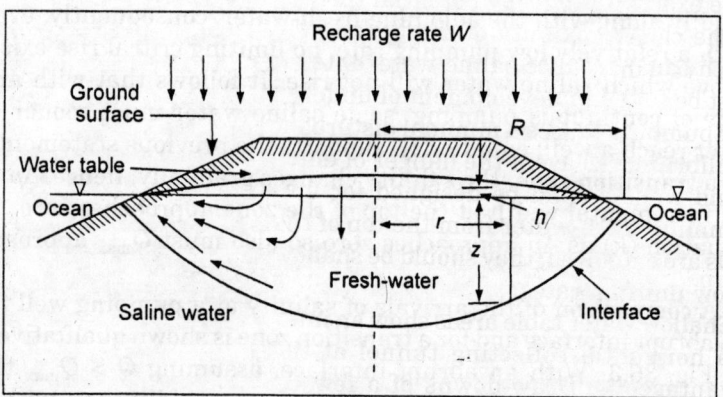

Fig. 36.4: *Fresh-water lens in an oceanic island under natural conditions.*

From the Ghyben-Herzberg relation and the Dupuit assumption an approximate fresh-water boundary can be determined. Assume a circular island of radius R, receiving an effective recharge from rainfall at a rate W. The outward flow Q at radius r is

$$Q = 2\pi rK\left(z + h\right)\frac{dh}{dr}$$

where,
K is the hydraulic conductivity and h and z are defined in Fig. 36.4,
$h = (\Delta p/p)z$ and that from continuity
$Q = \pi r^2 W$, then

$$z\,dz = \frac{Wr\,dr}{2K\left(1 + \dfrac{\Delta p}{p}\right)\left(\dfrac{\Delta p}{p}\right)}$$

Integrating and applying the boundary condition that $h = 0$ when $r = R$,

$$z^2 = \frac{Wr\left(R^2 - r^2\right)}{2K\left(1 + \dfrac{\Delta p}{p}\right)\left(\dfrac{\Delta p}{p}\right)}$$

Therefore, we can say that the depth to salt water at any location is a function of the rainfall recharge, the size of the island, and the hydraulic conductivity. For almost all island conditions, it can be shown that this approximate solution is indistinguishable from more exact solutions by potential theory.

The close proximity of the boundary zone to the water table which can introduce saline water into a well by upcoming. Care must be taken in development of underground water supplies, so that pumping causes a minimal disturbance to the fresh-salt water equilibrium. To avoid the danger of entrainment of saline water, island wells should be designed for minimum draw-down, just skimming fresh-water from the top of the lens. If small-diameter wells are employed, they should be shallow, dispersed, and pumped at low unifrom rates.

Shallow water table areas show an infiltration gallery consisting of a horizontal collecting tunnel at the water table, which is advantageous. Draw-downs of a few centimetres can in many instances furnish plentiful water supplies. Installations of infiltration galleries for local water supplies exist on Bermuda and the Bahamas in the Atlantic ocean and on the Gilbert and Marianas islands in the Pacific ocean. Where water tables are deep, dug wells or shafts are sunk to the water table with horizontal tunnels extending outward to intercept the uppermost layer of fresh-water.

The occurrence of fresh-water on an oceanic island is Barbados. Here a thin highly permeable coral limestone layer serves as the aquifer. Recharge from rainfall percolates through solution channels along the bottom of the limestone until it reaches sea level. From there to the coast the fresh-water floats in a layer (as sheet water) above the underlying saline water. Water is extracted from large-diameter dug wells connected to horizontal tunnels at the water table.

The balance between fresh and saline ground water on an oceanic island is Honolulu. Permeable basalt forms the aquifer, but impermeable caprock acts as a ground water dam. Before wells were first drilled in 1880, ground water discharged as spring either at the inland or at submarine boundaries of the caprock. At that time the transition zone was narrow and nearly horizontal. Subsequent development of water by wells has lowered the water table and expanded the transition zone so that the volume of fresh-water has been substantially reduced.

SEAWATER INTRUSION IN KARST TERRAINS

Aquifers of coastal regions consisting of karstic limestone pose special problems of seawater intrusion. Irregular fissures and solution openings enable seawater to enter the aquifer in configurations that may differ appreciably from those for more homogeneous aquifers. Unique features sometimes found in karst are intermittent brackish springs, which can result from channels connecting with the sea or where saline heads under high tides exceed inland fresh-water heads, causing seawater to be discharged inland.

The karst regions surrounding the Mediterranean sea are solutions channels, which discharge fresh-water as submarine springs. But pumping of these channels to prevent wastage of fresh-water often yields saline water within hours because seawater can freely enter the cannel if the fresh-water flow is reduced. To overcome this problem a two part dam shown in Fig. 36.5, which

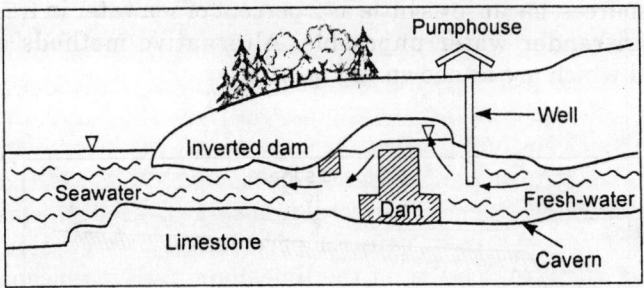

Fig. 36.5: *Schematic diagram of an underground dam in a coastal limestone cavern to prevent mixing of fresh and saline water.*

generally prevent the entry of seawater into a pumping well located upstream. This technique is successfully tested in the limestone caverns of many places of the world.

CONTROL OF SALINE WATER INTRUSION

For controlling saline water intrusion, knowledge of source of the saline water, the extent of intrusion, local geology, water use, and economic factors are necessary. Table 36.1 summarises the generally recognised methods for controlling intrusion from

Table 36.1: Methods for controlling saline water intrusion

Source or cause of intrusion	Control methods
Seawater in coastal aquifer	Modification of pumping pattern
	Artificial recharge
	Extraction barrier
	Injection barrier
	Subsurface barrier
Upcoming	Modification of pumping pattern
	Saline scavenger wells
Oil field brine	Elimination of surface disposal
	Injection wells
	Plugging of abandoned wells
Defective well casings	Plugging of faulty wells
Surface infiltration	Elimination of source
Saline water zones in fresh-water aquifers	Relocation and redesign of wells

various sources, bacause as little as 2 percent of seawater in fresh-water can render water unpotable, Alternative methods are discussed which are as shown in Fig. 36.6.

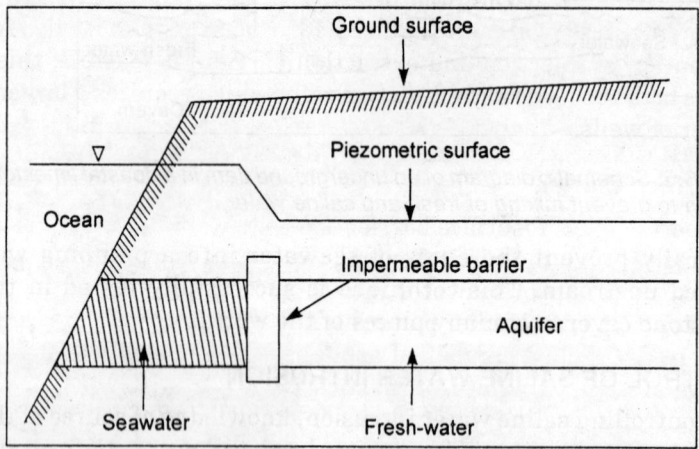

Fig. 36.6: *Control of seawater intrusion by an impermeable subsurface barrier paralleling the coast.*

Modification of Pumping Pattern

The locations changing of pumping wells, typically by dispersing them in inland areas, can aid in re-establishing a stronger seaward hydraulic gradient. Also, reduction in pumping of existing wells can control the saline water intrusion.

Artificial Recharge

Artificial recharge maintain and raise the ground water levels by using surface spreading for unconfined aquifers and recharge wells for confined aquifers. This necessitates development of a supplemental water source.

Barrier of Extraction

An extraction barrier is created by maintaining a continuous pumping trough with a line of wells adjacent to the sea. Seawater flows inland from the ocean to the trough, while fresh-water within the basin flows seaward toward the trough, as shown in Fig. 36.6. The water pumped is saline and normally is discharged into the sea.

Injection Barrier

Injected fresh-water flows both seaward and landward. High-quality imported water is required for recharge into wells. A combination of injection and extraction barriers is feasible; this reduces both recharge and extraction rates which requires a larger number of wells.

Subsurface Barrier

An impermeable subsurface barrier is constructed parallel to the coast and through the vertical extent of the aquifer can effectively prevent the inflow of seawater into the basin (as shown in Fig. 36.5). To construct a barrier materials needed might include sheet piling, puddled clay, emulsified asphalt, cement grout, betonite, silica gel, calcium acrylate, or plastics, etc.

Long Island, New York

A wedge-shaped mass of unconsolidated sediments underlain the long is land that extends more than 600 m in depth. Major intrusion has occurred in west. Long island as a result of development of ground water and of a decrease in recharge by improved drainage and sewer systems. By the mid 1930s pumping had lowered water levels at the western end to as much as 10 m below sea level and caused extensive intrusion. The southern coast similar development and reduction of recharge are causing active intrusion at depth.

Abandoning pumpuing cantrolled the intrusion and using an imported water supply and by requiring that water pumped from industrial wells be recharged back into the ground after use. To prevent this tendency an extensive program to salvage storm runoff and recharge it through infiltration basins has been initiated. In addition, experimental work is underway to reclaim wastewater for recharge to form an injection barrier.

RECOGNITION OF SEAWATER IN GROUND WATER

Ground water samples collected in zones of seawater intrusion after analysis show a chemical composition differing from a simple proportional mixing of seawater and ground water. Modifications in composition of seawater entering an aquifer can occur by three processes: (a) base exchange between the water and the minerals of the aquifer, (b) sulphate reduction and substitution of carbonic or other weak acid radicals, and (c) solution and precipitation. Only (C) process can change the total salt concentration; however, the first two processes, which require maintenance of ionic balance, can alter the percentage by weight of different salt components and thereby the total dissolved solids in milligrams per litre. Revelle recommended the chloride-bicabonate ratio as a criterion to evaluate intrusion.

Actually, the $Cl/(CO_3 + HCO_3)$ ratio is employed for practical purposes. Chloride is the dominant anion of ocean water, which is unaffected by the above processes, and normally occurs in only small amounts in ground water. On the other hand, bicarbonate is usually the most abundant anion in ground water and occurs in only minor amounts in seawater.

Drilling

Drilling work are to be carried out with the intent of maximising the chances of success in completing wells to the target depths. The beginning of operations at a particular location, reaming or installation of casing and beginning and completion of all cementing operations and whenever the operations become inappropriate due to a change in the drilling conditions or due to a change in the immediate objectives of the operation and at any time when a drilling problem threatens the well. For example, loss of circulation and caving formations.

Reaming of wells is nothing but to increase the borehole diameter.

Verticality of Wells

It is tested by the plum-bob method. The dummy consist of an axially suspended cylinder with a diameter twenty-five millimetres less than the hole size and three metres long. The suspending wire are less than five millimetres diameter of uniform cross-section with no kinks. Deviation must not be more than 10% and the reading of deviation and direction are generally taken at three metres depth intervals. It is to be noted that where deviation exceeds 10% the well may be deemed lost.

Circulation Fluids

For the completion of drilling wells, water quality must be under permissible limit. No saline water can be used. *Stiff foam* drilling fluids are used in drilling or reaming work by rotary rigs. It is done to enhance the quality of the formation sampling by limiting cavity and wash. It is to allow more precise detection of the occurrence, yield and quality of water. It is important to cave foaming agents must be biodegradable anionic surfactants.

Drilling Mud

When rotary mud drilling is required, the condition of mud (bentonite + water, mixed in 4 × 4 m × 6 feet deep pit) should be as follows:

Weight

Less than one point two kg/litre until artesian aquifers are encountered. Baryte may also be used for mud weight control.

Marsh Funnel Viscosity

Generally, less than 50 seconds, viscosity maintained. High water yield are with loose formation, the viscosity should be approximately 45 seconds. Generally, we notice that viscosity of mud pit is of higher value than flow line. But in more clayey formation, it is found that flow line viscosity is higher than mud pit, viscosity.

Sand Content

Sand content in bentonite mud should be less than one percent and in no case more than 2 percent.

Mixture of high-quantity bentonite equivalent to Wyoming bentonite in conjunction with cellulose polymers (e.g. PAC–RCMC) or inhibited biopolymer. (e.g. Kelzon XC polymer EZ mud). It is of international interest that when bentonite mud is used, it should be mixed and left to hydrate at least 12 hours before being pumped down the borehole.

To control the flowing wells, mud (Bentonite) weight should be around zero point two pounds (0.2 pounds) above the balanced

Fig. 37.1: *Mud drilling continuing in Oman.*

Fig. 37.2: *Drilling pipes installation during drilling process in Oman.*

mud weight. Sealing the artesian aquifer using casing or cementing techniques.

Lost Circulation Zones

For the lost circulation zone we do the following steps.
1. The installation of casing in specific portions of the well with the intent of sealing the lost circulation zone.
2. Cementing in short sequences to control weak formations.
3. The use of *blind drilling* techniques to obtains some drilling progress prior to the installation of the casing.
4. Use of geophysical logs to check the leak zones.

Downhole Cementing

It should be sulphate resistant Portland cement. The slurry (cement + water) made in 1×1 m diameter tank and must be of specific gravity from 1.68 to 1.8 with total dissolved solids (TDS) less than 1200 mg/litre and sulphate is less than 200 mg/litre. Let us see this.

How to calculate the amount of cement required to put down the hole in water well.

If the drilled hole diameter is 12.25 inches, we convert to volume per linear metre which is as follow.

$$\frac{12.25'' \times 12.25''}{1973.525} = 0.076 \text{ m}^3 \text{ per linear metre}$$

For 4 metres of annulars
$0.076 \times 4 = 0.304 \text{ m}^3$ slurry
One bag or sack of cement gives 50.18 litres of slurry

$$\frac{304 \text{ litres of slurry}}{50.18 \text{ litres of slurry bag}} = 6 \text{ sacks or bags of cement}$$

Each sack or bag of cement required 34.31 litres of water.

Note: To get specific gravity of cement of approximately between (1.68 × 1.8) which is used down the hole, we need to mix 6 bags or sacks of cement with 206 litres of water to get 304 litres of slurry in order to cover 4 metres of 13 $^3/_8$" diameter hole of theoretical volume.

Sampling Procedure

Representative samples of the strata penetrated will be collected every metre.

A sample of the formation cuttings are removed from the drilling medium by collecting the sample in a screen. It is necessary

Fig. 37.3: *Cleaning the borehole in Oman.*

Fig. 37.4: *Geophysical logging after drilled borehole in Oman.*

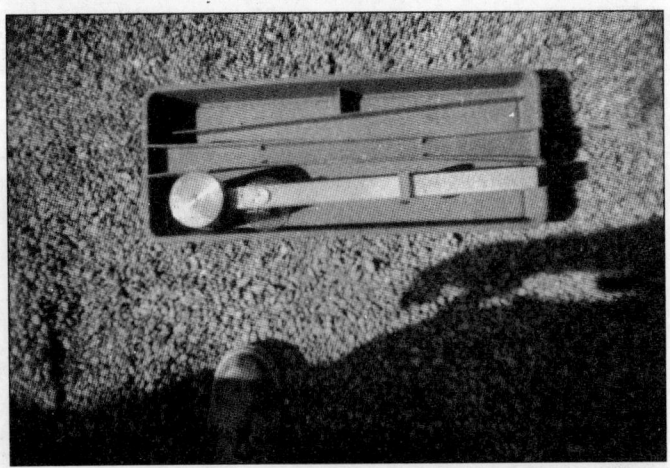

Fig. 37.5: *Measuring Sp. gr. of downhole cement in Oman.*

to mention here that are must be taken to ensure that the sample is representative of the material being drilled and should not be contaminated by hole erosion or cavings.

A container represents the sample box fitted with individual compartment along with the sample, indicating in waterproof ink

Fig. 37.6: *Author in brown uniform checking viscosity of bentonite mud in Oman.*

the depth from which the sample was recovered, and samples are generally placed in approximately marked heavy plastic sample bags, so that the entire geological section is clearly represented.

Fig. 37.7: *Borehole completed and ready for well head and cap fixing in Oman.*

Well Heads

The ends of the surface casing and well casing are usually cut-off to the lengths required. The annular space between the surface casing and well casing are generally backfilled. After this, the uppermost two metres of annular space are filled with concrete and neatly finished at the top.

An external well cap is provided and fitted but it is most important thing to care that the cap is closely fitted and fully seated on the top of the casing and the locking bladed are fitted so that it can be easily inserted and removed.

Fig. 37.8: *Well no. EB-04 fully completed in East Batinah over to Ministry of Water Resources of Sultanate of Oman.*

Gabion

Gabion construction is usually made to prevent flood damage of the well-heads. It is constructed before the casting of the concrete block around the casing the area around are excavated to a depth of one metre and a width of two metres to accommodate the bottom gabions. The bottom gabions are then be placed in position.

Concrete

Concrete should be a mixture of Portland cement and aggregates made up of at least five sacks or bags of 50 kg each per cubic metre of finished concrete. The mix should not contain large amounts of fine materials and mixing water are generally fresh, potable water and aggregates must be sound, durable and well-graded in sized ranging from coarse sands to rock of four centimeters to eight centimetres in size. Mixing must be done is mixing boxes and not directly on the ground.

V-Notch Weir Board

The weir board must be vertical equipped with a suitable graduated gauge which are installed to enable accurate reading of the volume of water being pumped. The flow over the weir must be in free fall.

Test Pumping Operations

The step drawdown test normally consists of five different discharge rates each of 60 minutes duration or any duration. The test should begin with the lowest discharge rate and increase consecutively until the maximum discharge rate is reached upon completion of the step drawdown test; a step recovery test normally lasts, for 5 hours.

After step recovery test completion, the constant discharge pumping test begin or it may be carried out on the day following the step drawdawn test, however, the time between the tests which allows for the recovery of the water levels in the wells will be 24 hours. During the period of the test in presence of well site geologist, the measurement and record water levels in the observation wells are done.

The recovery test begins immediately on cessation of the constant rate discharge test. The water level recovery period generally continues until the well has recovered to within 95% of the total drawdown. It is generally in practice that measurement and record water recovery levels in the pumped well and in the observation wells after the pump is shutdown. It is very important to care that pump should not be removed from the well until the completion of the recovery test.

Requirements at Drilling Site

Safety is the prime need of water well-drilling. An officer must be present during working hours at well-drilling site to fulfil safety requirements on drilling and test pumping operations. This officer is called by the name of *Safety Officer*. He must be able to render effective emergency first aid and cardiopulmonary resuscitation when required to do so. For this, a well-maintained, comprehensive first aid kit is to be kept at each work site and the camps at all times.

The safety officer is responsible for all aspects of safety on site and remove as many hazards as possible and equipment which is required for safe drive of works at water well-drilling site.

Personnel working at the location must have protective glasses, ear protection when noise levels exceeds 85 decibels, safety boots, rubber boots, gloves, rubber aprons, rubber gloves, face masks and overall two spare sets of personnel protective equipment must be provided at the site for use as by authorised visitors.

First Aid Equipment

First aid equipment comprises a comprehensives first aid box, an emergency shower and eye wash facility and sufficient stretchers to cater for an emergency.

Fire control equipment in good workable condition and maintained regularly. Work area should have nine litres type water extinguisher and twelve kilogram dry powder extinguishers.

Gas detection and protection equipment during operation, noxious or explosive gases like methane (CH_4) or hydrogen sulphide (H_2S) may be encountered. For this gas detection and protection equipment must be provided at the drilling site. This includes wind socks, breathing apparatus, personal H_2S monitors, a testing unit for personal H_2S monitors, a portable multigas detector complete with hand-held probes and an aspirator pump for hydrocarbon gas, hydrogen sulphide and oxygen.

The drilling rig, mud pumps, air compressors, generators, vehicles, etc. must be free of exposed moving components and have emergency shutdown procedures. Last but not least base camp and mobiles in remote areas where communications are considered necessary must be available for safety reasons.

Drilling Engineer

Drilling engineer plays an important role while drilling is going on. He anticipates drilling and test pumping problems and to plan timely solutions. A drilling engineer is also responsible to ensure

Fig. 37.9: *Author as a geologist in the chair with drilling rig, completed well and some of drilling crew members in Oman.*

Fig. 37.10: *Author in uniform relaxing with Omani geologists during break at site in Oman.*

quality control and properly engineered operations during all the work, and also give orders to the drilling superintendent and/or test pumping superintendent.

Fig. 37.11: *Author (as a geologist in Sultanate of Oman) studying the samples of well cuttings systematically to make geologic log.*

Fig. 37.12: *Author (as a geologist, with red) with crew before gravel pack in Oman.*

A qualified drilling engineer must have sufficient experience in the supervision of exploration and ground water drilling projects. He must be familiar with the use of polymers and bentonite mud. Stiff foam and air foam as circulation fluids and well-known about pressure cementing procedure.

Well Site Geologist

The well site geologist will remain at the drilling location during all periods of drilling operation. The geologist should be equipped with 7 to 40 times binocular microscope, an angle poise lamp and a hand held or vehicle mounted global positioning system. He must be well qualified and have sufficient experience in gathering well site hydrogeological data, drilling cuttings description, hydrogeological techniques and also must be able to interpret the results of downhole geophysics.

Geological Considerations During Water Well-Drilling

It is the duty of a wellsite geologist to make it sure that lithological samples are collected at one metre intervals, and apart from those that are to be packed and dispatched and for inspection each sample is laid out in a logical order. The depth of each should be marked. The samples should also be clean in such a way that clays, if present are not washed out. Intervals at which water strikes are suspected should be reported accordingly and samples to be retained are marked clearly with borehole number and depth interval.

The geological field log should be maintained up-to-date at all times and that the composite borehole log is updated on a daily basis. The latter may have to be done during the evenings. The composite borehole log is a fair copy from which somebody else may be transferring data to a computer programme, therefore it should be neat and clean.

Hydrogeological Considerations During Well-Drilling

A hydrogeologist records pH, EC of make-up water if it is not from the current borehole. He repeats these tests with each new bowser and record on "Wellsite Physicochemical Monitoring Record".

After the first water strike a hydrogeologist records V-Notch reading, both in height and discharge continuously, record of pH, (Eh in–mv) EC (electrical conductivity in cm/sc.) and temperature. At every one metre interval and the full suite of tests whenever the V-notch reading indicates that further waters have been encountered. Then he makes notes on the geologic field log, making the correlation between water strikes/lithology/penetration rates at the time of the water strike.

A hydrogeologist also records water level at the beginning and end of every shift and after any extended period of the rig standing mentioning date and time then measures water level at the cessation of airlifting, allow 20 minutes recovery and monitors again.

A hydrogeologist must be familiar with the new laboratory procedures for calibration and operation of the DO (dissolved oxygen) pH, Eh and EC metres and alkalinity titrations.

On each borehole completion one complete set of these forms should be handed over to project office.

It is to be noted by a wellsite geologist that the guidelines for rock descriptions. As set out by the geology section, should be followed. Here we have some examples of common lithology descriptions below as a case study of a monitoring well-drilled in Sultanate of Oman.

Well No.	-	EB – 03 (R)
Locality name	-	Bani Kharus Abu Abali
UTM Coordinates	-	05 – 69 – 325 E 26 – 14 – 152 N
Start date	-	20.01.2004
Date of completion	-	25.04.2004
Monitoring well	-	8½″ diameter hole
First water strike (FWS)	-	At 52 m
Main water strike (MWS)	-	At 56 m
Drilling method	-	Rotary, air foam flush (0 – 65 m) mud drilling (65 – 432 m)
Casing type and depth	-	13 $^3/_8$″ ID mild steel surface casing + 0.30 – 4.9 m 5½″ steel casing + 0.55 – 202.7 m depth. 5½″ to 3½″ steel crossover 202.7 – 203.2 m depth 3½″ steel screen 203.2 – 316.69 m depth
Total depth of borehole	-	432 m
Method of well development	-	Air lift well development with compressed air

Lithology/Formation of Well EB-03 (R)

0 – 2 m : *Sandy gravels*: Pale brown fine sands (70%), mixed with gravels of dark brown ophiolites (15%), and grey limestone (15%).

2 – 34 m : *Pebbles mixed with coarse sands*: Rusty brown ophiolites (55%) and whitish-greyish limestone (35%). Here cementing materials also found in scattered way as up to (08%) + red chert (01%). In some samples minor quartz also reported. Grain size increases as gravels from 12 m onwards with 60% limestone and 40% brown ophiolites. Formation is somewhat fresh. From 31–34 m, pebbles of normal weathered rusty brown ophiolites (80%) with 20% whitish-greyish limestone.

34 – 46 m : *Loose gravels*: Brownish ophiolites (60%) which are normally as rusty brown in colour mixed with changed composition as grey limestone (35%) and quartz (05%). Limestone gravels percentage goes on increasing from 37–46 m as up to (65%) with 30% brownish ophiolites and quartz (05%).

46 – 56 m : *Calcrete*: Brownish, pinkish, most friable calcrete (70%) mixed with greyish-whitish limestone (15%) + dark brown ophiolites gravels (15%). Formation shows somewhere slightly or partially cementation and somewhere slightly weathered.

56 – 82 m : *Loose gravels*: Colourful formation as rusty brown ophiolites (45%) mixed with whitish-grayish limestone (35%) + brownish-pinkish calcrete (15%) + calcite (05%). There is almost no calcrete at 57 m. About 10% silt and clay are reported at 61, 62, 65, 78 and 79 m depth.

82 – 107 m : *Silt and clay*: Silt and clay are whitish-pale brown in colour which contain 88% of the whole sample with rusty brown ophiolites (08%) and (04%) is grey limestone pebbles, particularly at depths 83–85 m. Further clay percentage decrease little bit. Somewhere some specks of calcrete also no-

ticed. From 104 to 107 m, very coarse sands of grey limestone (55%) + brown ophiolites (30%) + quartz (03%) + silt and clay (12%).

107 – 150 m : *Loose gravels*: Gravels of whitish-greyish limestone (60%) rusty brown ophiolites (25%) + quartz (04%) + magnesite (03%) + calcite (03%). There are minor clay also recorded which is about (05%). More brownish clay are reported from depths 126–128 m and 132–135 m. Slightly weathered ophiolites also noticed. Grain size decreased 138 m onwards as coarse sands.

150–175 m : *Pebbly calcrete*: Whitish, pinkish and buff coloured calcrete contains about 40% of the sample. Pebbles of limestone which are greyish-whitish in colour constitutes (35%). Brownish pebbles of ophiolites (20%) and red chert (05%).

175–184 m : *Pebbles*: Pebbles of whitish-greyish limestone (60%), greenish-brownish ophiolites (25%), pinkish-whitish calcrete (10%) + red chert (05%) at 181–183 m, mior clay also reported.

184–315 m : *Sandy pebbly gravelly calcrete*: Greyish-whitish limestone (50%) + whitish-pinkish calcrete (30%) + rusty brown ophiolites (14%) + quartz (03%) + chert (03%). From 250–275 m calcrate percentage increased up to (45%) silt and clay (05%) also reported. From 275–288 m more greyish limestone (60%) + calcete pebbles (25%) + brownish ophiolites (10%) + calcite (02%) + quartz (02%) + chert (01%). Calcrete of pebbles size increase in percentage up to (55%) + greyish limestone (30%) + brownish ophiolites (10%) + quartz (03%) and chert (02%). This is the case of 300–315 m depth.

315–363 m : *Silty clayey sands*: Whitish, creamy-brownish silt and clay (75%) + whitish-pinkish, calcrete (15%) + greyish limestone sands (08%) + quartz (02%). From 321–336 m, silt and clay percentage increased up to (80%) and 336 m onwards silt and clay percentage decreased to (50%). From 348–353 m limestone is more about (40%) with slightly weathered with a few of them are also partially cemented brown ophiolites (25%) and clay percentage goes down to (30%).

363 – 368 m : *Sandy calcrete*: Coarse sands of limestone (45%) which is grey in colour. Which calcrete (35%) + brownish ophiolities (15%) + chert (05%), minor clay also reported somewhere.

368 – 432 m : *Silty clayey coarse sands*: Whitish, creamy, brownish silt and clay (80%) + calcrete (whitish) about (08%) + coarse sands of grey limestone (07%) + brownish ophiolites (05%). At 375 and 376 m pure MARL is recorded.

Total depth = 432 m.

Total casing depth = 317.09 m.

Steel blanks (casing) of 5½″ diameter = + 0.55 – 202.60 m.

Steel crossover 5½ to 3½″ diameter = 202.60 – 203.10 m.

Steel screens of 3½″ diameter = 203.10–317.09 m.

Air Lift Well Development with Compressed Air "At 244.80 m Depth"

Time of development: 16:30 to 19:30 hrs = 3 hours.

Water quality check , we check the quality of water after well development. We do development until muddy water or dirty water become clean and clear.

Electrical conductivity in ms/cm = 265 ms/cm

Temperature in °C = 36.9 °C

Hydrogen ion concentration (pH) = 8.12

Eh in (mv) = –72.2 mv

Yield (q) = 18.45 l/s

At 310.80 m Depth

Time of well development: 20.00 to 22.30 hrs – 2½ hrs

Electrical conductivity in µs/cm = 268 µs/cm

Temperature in °C = 35.1 °C

Hydrogen ion concentration (pH) = 8.09

Eh in (–mv) = – 73.1 mv

Yield (q) = 19.50 l/s

Total Air Lift Well-Development with Compressed Air of Well No. EB-03(R)

3 hrs + 2½ hrs = 5½ hrs

(Completion of well no. EB-03® is over)

Thus, we may conclude by summarizing the daily responsibilities of a hydrogeologist at water well-drilling site. After surface casing setup and drilling will begin. He fills in form with well name, date, bit size and name, depth of surface casing, diameter and information on drilling fluid for hole drilled below surface casing. While drilling, record, penetration rate, slow drilling, etc. Record of lithological intervals drilled and total depth drilled that day, time, date. It water has been struck, i.e. first water strike (FWS), any other zones yielding water. Then record, V-Notch weir reading, airlift yield, EC, T, PH, Eh, DO, etc. He also plots airlift, EC and penetration while drilling.

A hydrogeologist record static water level each day, record time that drilling resumes. He also assists the loggers in locating site. If it is necessary he brings logs back to office and well designing and lastly he collects a water sample at the end of development. So, it is the duty of a geologist to make assure that all drill logs are complete and it is final to be filed.

Drilling Superintendent

The duty of the drilling superintendent to be responsible for the day-to-day operation of the drilling operations. He should have sufficient experience in supervising the drilling and completion of wells and must be familiar with the use of polymers and bentonite muds, stiff foam an air foam as circulating fluids and be familiar with pressure cementing procedures.

Test Pumping Superintendent

The superintendent of test pumping will remain at the site to provide liaison with the engineer on all matters related to the successful completion of the pumping tests. He has to ensure quality control and properly engineered operations during all of the work and he has also to anticipate operational problems and to plan timely solutions in consultation with the engineer.

The superintendent of test pumping must have extensive experience in the supervision of aquifer testing projects. He must be somewhat expert in the use of generators, submersible tests

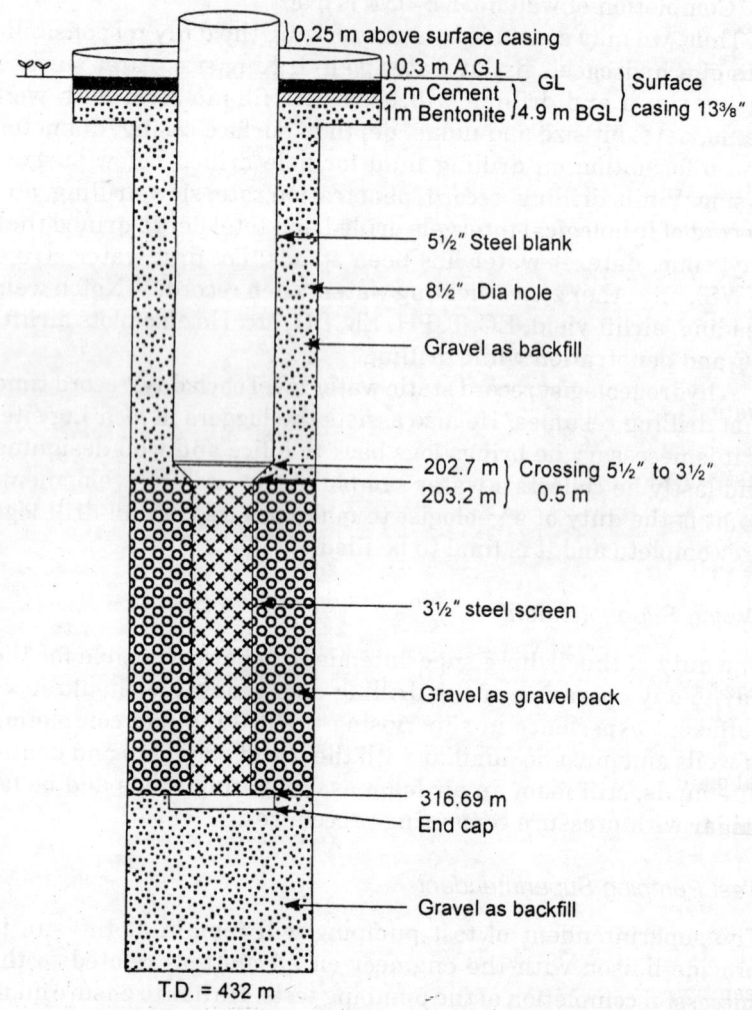

Fig. 37.13: *Construction of monitoring well.*

pumps and all associated measuring and recording equipment and operations in every type of environment of the site.

Borehole Completion Summary

Borehole number	:	21/9A
Contract number	:	92–21
Project	:	Piezometer drilling-A1 Batinah
Drilling supervision	:	Ministry of Water Resources

Location

Area	:	Wadi Bani Kharus, Eastern Batinah
Map reference	:	Series K-6611, Sheet No. NF- 40 3B (BARKA), 1: 100,000
Grid reference	:	573336.87 E, 2612397.74 N
Site ID	:	EM 713233 AA
Elevation (Hilti Nail)	:	58.766 mamsl

Construction

Date drilling started	:	02.05.93
Date well completed	:	26.05.93
Drilling method	:	Rotary, air-foam flush (0–246 m)/ Mud (246–TD)
Bit size and depth	:	17½″ Tricone roller bit 0–8 m
		8½″ bit 8 – 350 m
Total drilled depth	:	350 m bgl
Casing type and depth	:	$13^3/_8$″ ID mild steel surface casing + 0.4–7.3 m
	:	4″ ID PVC casing + 0.2–231 m
	:	1″ ID PVC casing + 0.2–81 m
Screen type and depth	:	4″ ID PVC screen at 219–227 m
		1″ ID PVC perforated at 76–81 m

Observations and Tests

Aquifer	:	Wadi alluvium
Water strikes	:	First strike 56 m, main strike 64 m
Static water level (m BMP)	:	4″–48.25, 1″–48.25 (10.06.93)
Original measuring point and elevation (before 11 Dec. 1993)	:	Top of steel surface casing = 58.785 mamsl
New measuring point and elevation (after 11 Dec. 1993)	:	Top of steel surface casing = 59.015 mamsl
Water quality (field measurements)	:	EC 450–644 µs/cm, pH 8.06–8.53
Maximum air-lift yield	:	40.7 1/s (226 m)
Aquifer tests	:	Nil
Geophysical logging	:	Hydrotechnical
		Fluid and formation logs 211.2 m (05.05.93)
		Fluid and formation logs 246 m (09.05.93)
		Full suite 350 m (13.05.93)
		TV borehole log 112 m (22.05.93)
Status	:	Completed as a twin piezometer observation well. However 4″ well has been abandoned because 4″ casing holed at 112 m BGL.

Test Pumping Crew and Support Staff

The test pumping crew commonly include a competent technician and test pumping hands capable of working on test pumping units operating to the heads and discharge. The crew must be sufficient in numbers to work 24 hours operations and be capable of undertaking regular monitoring of all pumping and observation wells.

The drilling and test pumping operations should be support with site based qualified and certified electricians. Welders and mechanics, licenced drivers for heavy and light vehicles.

A Case Study

Ministry of Water Resources	Composite borehole log	BH No. 21–9A

Contract no. 92/21 Contractor	Project piezometer drilling-al batinah	
Equipment T-4 W/Rotary	UTM Co-ords 573336.87 E 2612397.74 N Co-ords source Survey ELEV (mASL) 58.572 Source Depth referenced from Ground level	Site ID EM 713233 AA
Fluid Air foam/Mud		Field Wodi Bani Khar
Hole size 17½" to 8 m		Class Obs TDH 350 m
8½" to 350 m		Date started 02.05.93
		Date completed 26.05.93

Remarks: Rate Penetration rate (min/m), yield = air-lift yield (1/s) Maximum yield = 41.3 1/s (226 m)
  EC = Electrical conductivity (/cm)μS FWS = First water strike, MWS = Main Water Strike
  Status: Dual completion monitoring bore

1 \ 0 Rate 33 900 0 Yield 45	Description	Lith	BH	Const	Details	0 EC 700
10 20 30 40	GRAVEL: Gravels of ophiolite and limestone mixed proportion rusty brown weathered, grey, subrounded. Minor clay at 9–22.5 m. Minor coarse sand of ophiolite below 22.5 m.				17 ½" Borehole 13 3/8" ID Surface casing 12¼" Borehole 1" ID PVC casing to 81 m 4" ID PVC casing to 229 m Gravel to 229 m	
	CALCRETE: Brownish pink, calcareous, friable, slightly clayey below 34 m.				Gravel pack 47. 80 m (4") (10.06.93)	
50	GRAVEL SLIGHTLY CEMENTED: Gravels predominantly of limestone grey and black, minor ophiolite,		SWL 4	SWL 1	47.85 m (1") (10.06.93)	
60	GRAVEL: Gravels of completely weathered		MWS	FWS	First W Strike (56 m) Cement plug	
70	ophiolite, pale to pinkish brown, limestone few, rounded cobbles and pebbles.				(52–72 m) Main W Strike (64 m)	
80	CALCRETE: White to pinkish white, calc, hard.				Bentonite seal Sand pack	
90	GRAVEL SANDY: Gravels of limestone and ophiolite, granules, pebbles, rounded, well sorted. Minor sand.				1" ID PVC perfo (76–81 m) Sand pack Bentonite seal	
100	CALCRETE: Varigated colour, calcareous.					
110	GRAVEL: Gravels of ophiolite and limestone, rusty brown, grey. Slight comontation at 99–103 m.				Cement plug (85–105 m)	
120	Calcrete interbedded at 103–105 m.					
130	CALCRETE: Pale yellow to white, calcareous, slightly friable, clayey at 132 m. Gravel interbedded at 126.5 to 128 m.				8½" Borehole	
140	GRAVEL: Gravels of completely weathered ophiolite, brown. Minor clay at 146 m.				Gravel pack	
150	CALCRETE CLAYEY: White to pale pink, calcareous, clayey with few Mn stains.					
160 170	GRAVEL: Gravels of ophiolite, well to completely weathered					

Fig. 37.14: *Full description of log well.*

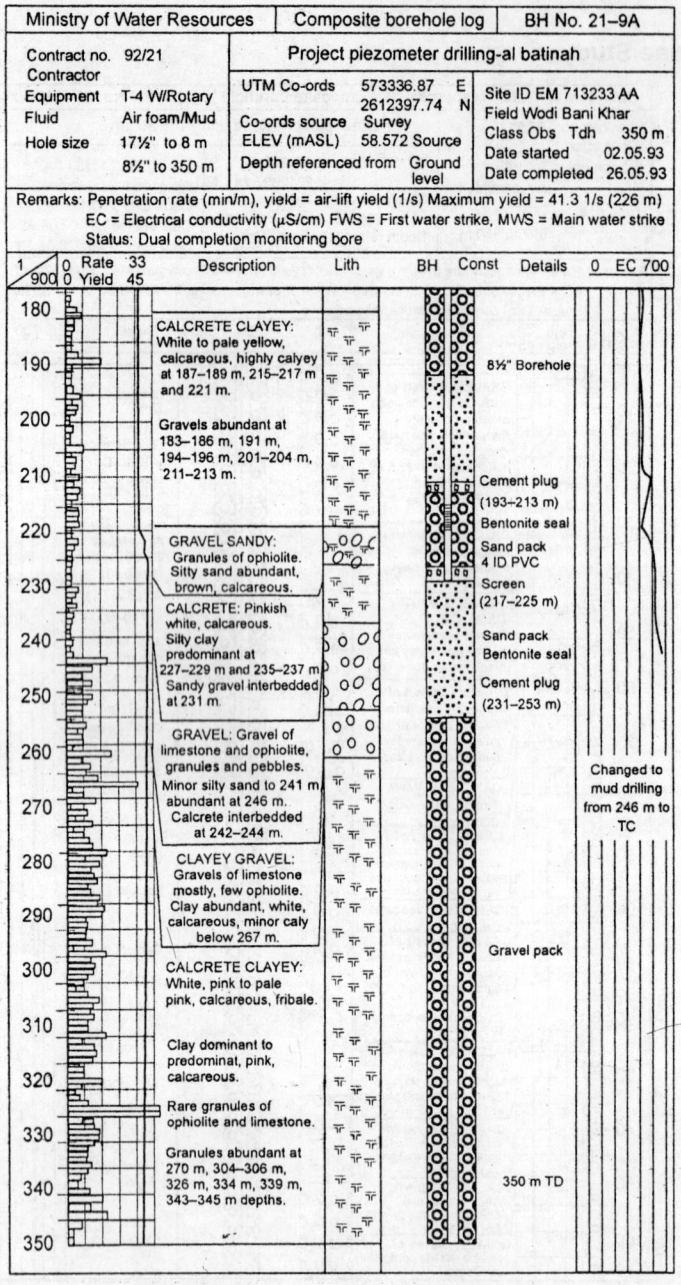

Fig. 37.14: *Full description of log well.*

Some Other Examples of Construction of Wells

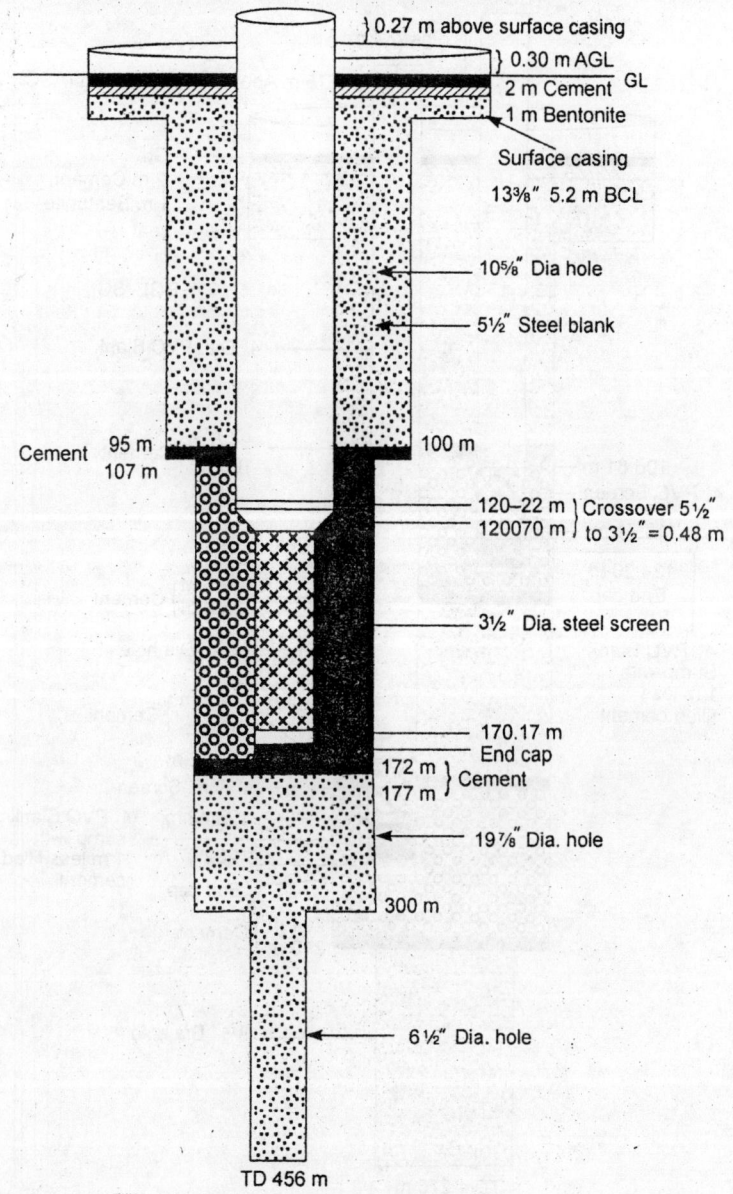

Fig. 37.15: *Monitoring well.*

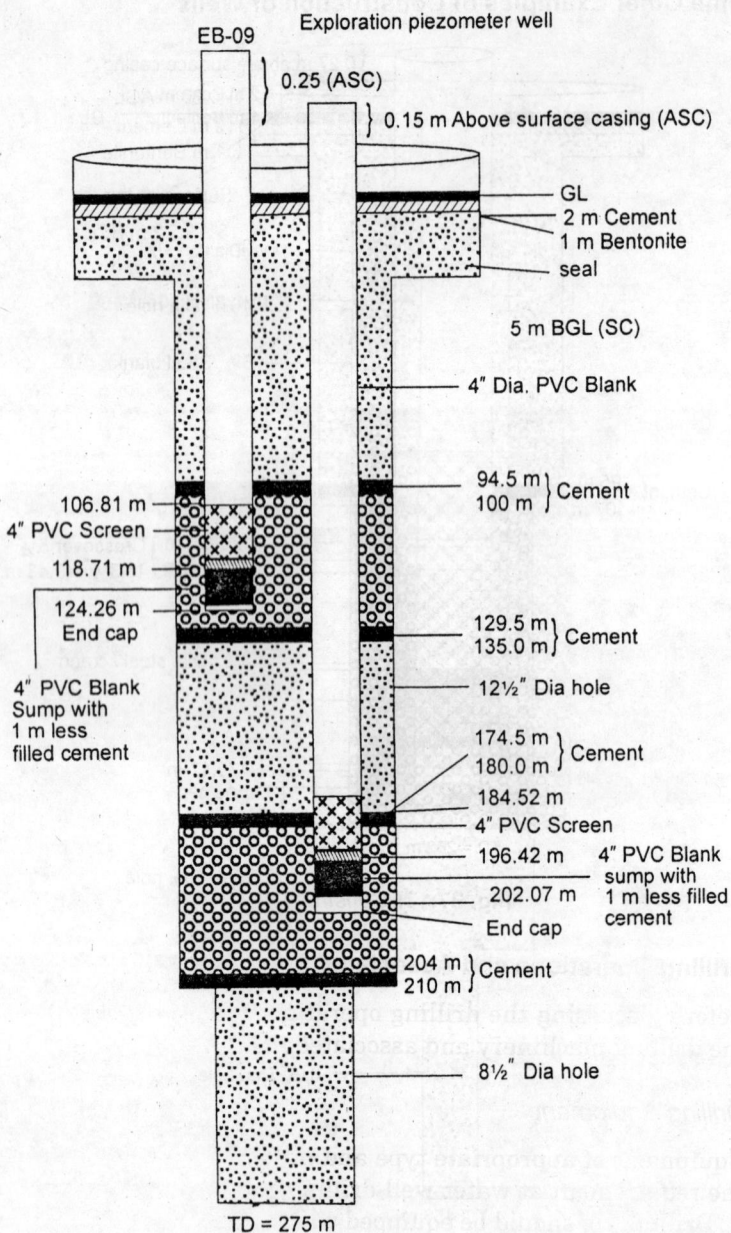

Fig. 37.16: *Dual completion of well.*

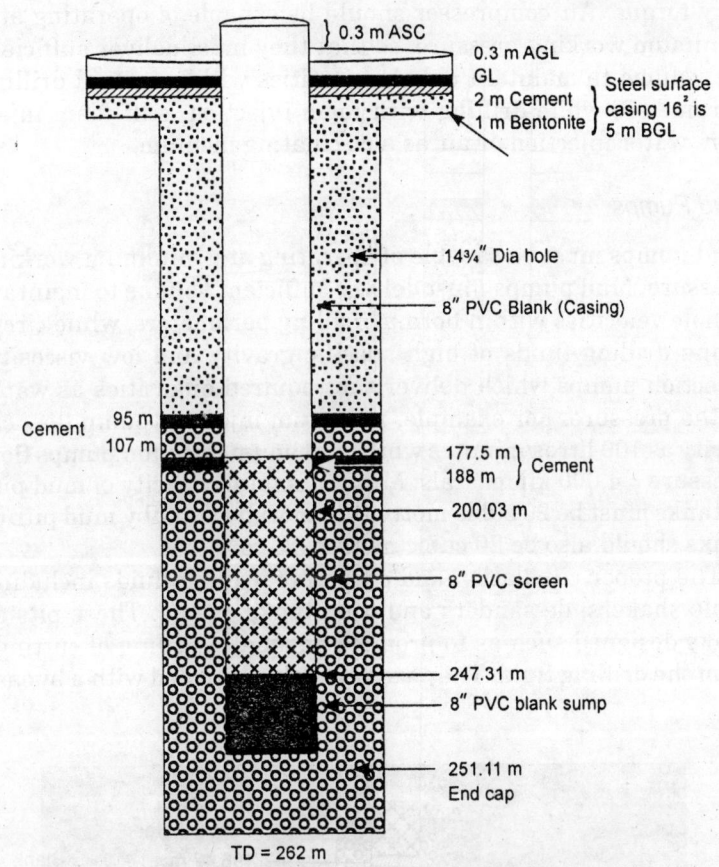

Fig. 37.17: *Construction log.*

Drilling Operations and Associated Equipment

Before discussing the drilling operation we must have a look at the drilling machinery and associated equipment.

Drilling Equipment

Equipment of appropriate type and in sufficient number to meet the requirement at water well-drilling site must be available.

Drilling rig should be equipped with necessary standard to operate safely will drill string or casing loads and to achieve a rotary torque. Air compressor should be capable of operating at a minimum working pressure, so that they must deliver sufficient air volume to maintain uphole velocities within normal drilling parameters while drilling with foam injection, stiff foam injection, water injection or air as a circulating medium.

Mud Pumps

Mud pumps must be capable of operating at a minimum working pressure. Mud pumps must deliver sufficient volume to maintain uphole velocities within normal drilling parameters, while circulating drilling fluids of high specific gravity and low viscosity. Injection pumps which deliver the required quantities as water at the pressure. For example, minimum injection pump flow capacity as 100 litres/minutes while maximum injection pumps flow pressure = 4,000 kilopascals. Minimum total capacity of mud pits or tanks must be 20 cubic metres (20 m^3) and standby mud pits or tanks should also be 20 cubic metres (20 m^3).

The proper circulation and control of drilling fluids including shale shakers, de-sanders and mud tanks or pits. These pits or tanks designed such as to promote adequate settling of cuttings from the drilling fluid. They are generally designed with a bypass

Fig. 37.18: *Drilling bits in Oman.*

Fig. 37.19: *Drum fixing before surface casing in Oman.*

facility which would allow rapid cleaning of contaminated sections of the tanks.

Tricone drill bits are hole openers to achieve the required depths and diameters. While giving good sample results. Drill collars, drill bits, calipers suitable for measuring ID and OD of all tubular. Stabilisers are generally of the welded spiral blade or roller type stabiliser which are suitable for all conventional well sizes for the maximum and minimum diameters.

Support equipment with the drilling rig to meet the needs during drilling operation comprise mainly a large capacity four wheel drive mechanical shovel which is generally used for road building and location preparation when moving between locations where no roads for easy transportant exist and also during mobilisation and demobilisation, etc.

To load and unload casing, drill pipe and drilling consumable such as mud and cement at the drilling location, four wheel drive lifting equipment with the capacity to lift and carry not less than three tones. Trucks and trailors to transport water, drilling equipment and consumable items to drill site.

Cementing Equipment

At each drilling rig, the 1 m³ (one cubic metre) volume of cement slurry usually mixed continuously. Centralisers are spring steel streps welded to hinged steel collars which are easily clamped onto the well casing. Cement basket supports the required column of cement and casing cementing shoes are stab-in-type.

For airlift pumping tests, the equipment generally consists of an airlift assembly which is capable of producing a discharge of at least 30 litres/second and a rectangular weir tank constructed from steel plate and fitted with a V-notch weir and water level monitor, which is large enough to measure discharges up to one hundred and fifty litres/seconds.

Test Pumping Operations Equipment

Test pumping operations equipment mainly constitutes a "Test Pumping Unit". Downhole pumps of appropriate diameter as maximum head envisaged as

Electrical submersibles = 150 metres.

Maximum head electrical submersibles = 10 metres.

Maximum pump discharge rate = 40 litres/second.

Minimum pump discharge rate = 2 litres/second.

Discharge Metering System

A flow metre and an orifice-plate discharge metering system of sufficient capacity to meet the required maximum discharge rates are generally install. Valves and other means of throttling the water discharge down to several rates of discharge are also required.

The dip tube of 25 mm, i.e. 1 inch with smooth internal surfaces are generally installed in manner such that a water level indicator can pass through it freely to measure water levels. Orifice pipes and water metres suitable for the measurement of water discharge should also be attached with correct venture.

Water level indicators are of 200 metres OTT, generally used. Electrical conductivity (EC) metres WTW combined Ec/Temp. Metre LF96 or equavalent are generally used.

Eh metres and PH metres of Hanna catalogue no. H19025C or Hanna combined PH/Temperature metre catalogue no. H19025C are generally in use.

Thermometres indicating up to 100 degree celsius, stop watches showing 24 hours and lay-flat hose with maximum pump discharge rates with minimum distance, i.e. 500 metres are required at water well drilling site.

Fig. 37.20: *GI pipes in borehole in Oman.*

Fig. 37.21: *Mud drilling in operation in Oman.*

Pumping Tests

Pumping test of wells and boreholes are carried out to determine the performance characteristics of the well and to determine the hydraulic characteristics of the aquifer. The hydraulic characteristics of the aquifer is the most important. The hydraulic characteristics of the aquifer are, transmissivity (T) and storage coefficient (S) and if these characteristics are known, then certain predictions can be made. For example.

(a) The effect of pumping from new wells on the water levels in existing wells nearby.

(b) The drawdowns in a well at future times and at different pumping rates.

(c) To assess the total flow through an aquifer.

(d) To assess the rate of flow through an aquifer.

(e) Usable volume of water stored in aquifer.

What is a Pumping Test

A pumping test involves abstracting water from a well, usually at constant rate, and then measuring the change in water levels caused by the pumping. The water level is measured in the well that is pumped, and also in other wells nearby which are called *observation wells* or *piezometres*.

Pumping water from the ground causes the water levels in the wells to fall. The greatest decline in water level is in the pumped well. The decline in the observation wells is less, the further they are from the pumped well. In fact, a cone of depression is created, centred on the pumped well as shown in Figs 37.22 and 37.23.

The pumping test provides information that can be analysed mathematically using a variety of equations.

The test may last a few hours, often 24, or it may be several days. Sometimes a test lasts for several weeks or months.

Test Equipment

The most essential equipments required for a pumping test are as follows.

(i) Pump (and generator).

(ii) Rising main (pipe from to surface).

(iii) Control valve-to-control flow rate.

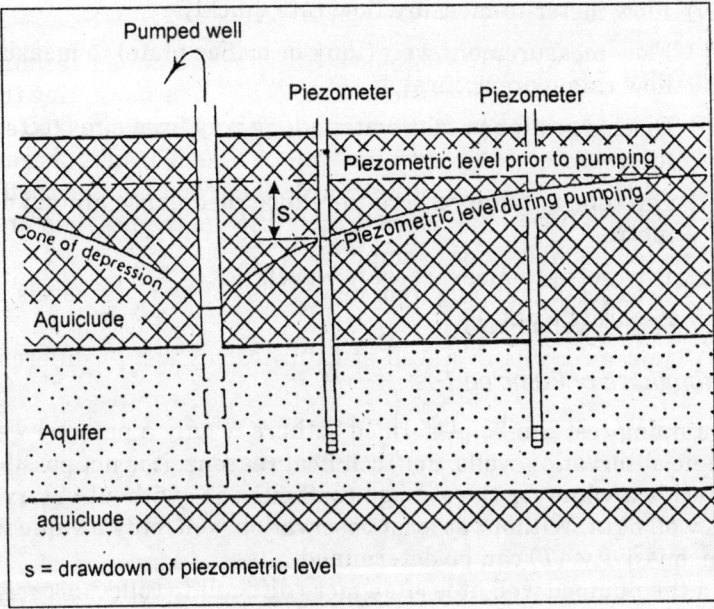

Fig. 37.22: *Drawdown in a pumped aquifer.*

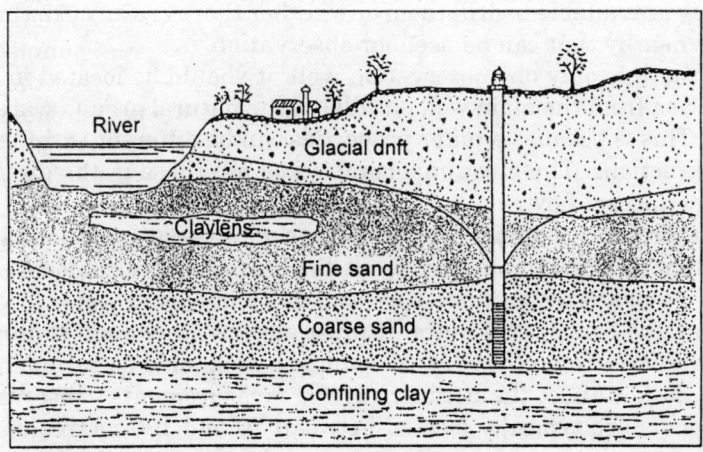

Fig. 37.23: *Cone of depression expanding beneath a riverbed creates a hydraulic gradient between the aquifer and river. This can result in induced recharge to the aquifer from the river.*

(iv) Flow meter-to-measure flow rate quickly.

(v) Flow measurement weir (tank or orifice plate) to measure flow rate more accurately.

(vi) Discharge pipe to take water a long way from site (100's of metres).

(vii) Dip tube in pumped bore-to-reduce turbulence around dip metre.

(viii) Dip metre-to-measure water level.

(ix) Accurate clock.

Performance of Pumping Test

A pumping test can be performed if there is only a pumped well and no observation wells. For technical reasons, it is not possible to derive a value for the storage coefficient (*S*) without observation well data. Without at least observation well, only a value for transmissivity, (*T*) can be determined.

In the pumped well, it is often more difficult to collect accurate water level data because of turbulence in the well caused by the pumping. But, to collect accurate water level data and to be able to estimate both *T* and *S*, it is preferable to have observation wells. The number of observation wells usually depends on how much money is available to drill them or whether there are any existing wells nearby that can be used for observation.

If there is only one observation well, it should be located in a direction from the main well parallel to the natural ground water flow. This is because aquifer properties can be different in different directions. The most important value are those in the direction of natural flow.

If there are a number of observation wells, these are located in different directions and at different distances. An observation well must be near enough to be sure that the changes in water level are not too small to measure. It must also be far enough so that interference from effects of the pumped well, such as turbulent flow, do not affect the accuracy of the measurements. Observation wells are typically located between 10 and 100 metres distance from the pumped well. One more important point to note that if there are other wells already existing nearby, such as farm wells, these should be used as observation wells, if possible.

Recording Drawdown

A dip metre is used to measure the water level in a well. This is a measuring tape with metre and centimeter marking on it. At the

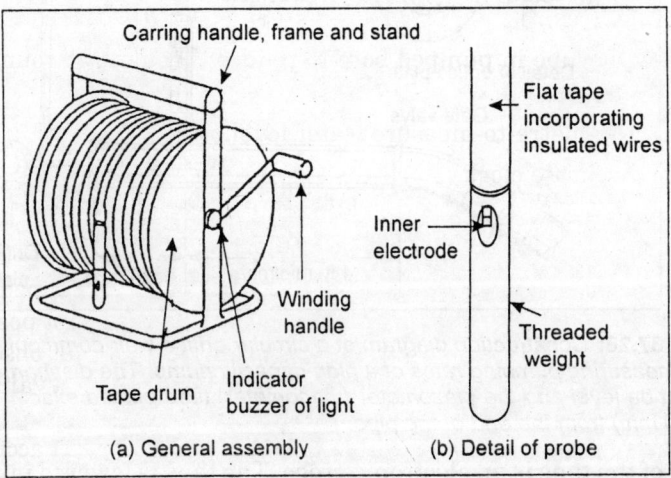

Carring handle, frame and stand

Flat tape incorporating insulated wires

Inner electrode

Threaded weight

Winding handle

Tape drum

Indicator buzzer of light

(a) General assembly

(b) Detail of probe

Fig. 37.24: *Most commercially available dippers are made on the same general format. Batteries (usually totalling 9 V) are housed in the drum spindle, which also contains the electronic circuitry. The probe is usually made of stainless steel or brass and acts as an electrode, with a second inner electrode being visible through holes in the weighted end. This weight is usually threaded and can be unscrewed to give access to the inner electrode.*

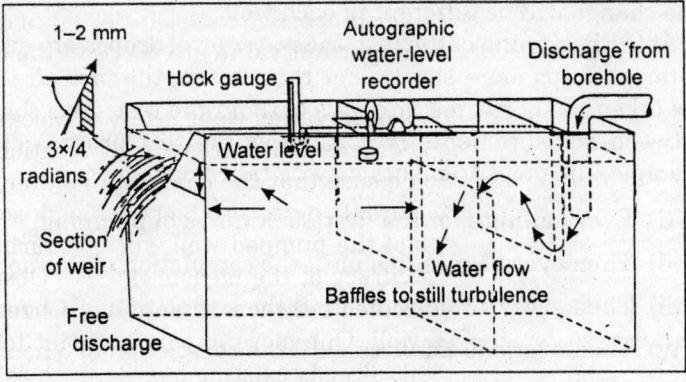

1–2 mm

Hock gauge

Autographic water-level recorder

Discharge from borehole

$3 \times /4$ radians

Water level

Section of weir

Water flow

Baffles to still turbulence

Free discharge

Fig. 37.25: *Weir tank.*

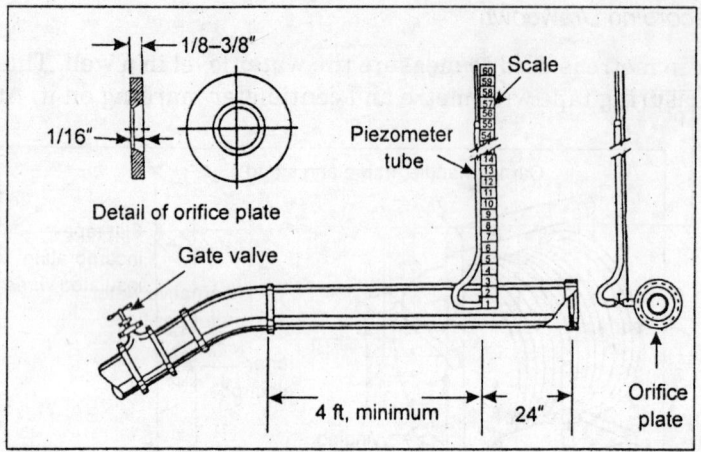

Fig. 37.26: *Construction diagram of a circular orifice weir commonly used for measuring pumping rates of a high-capacity pump. The discharge pipe must be level and the piezometer (manometer) tube placed exactly 24 in. (610 mm) from the end of the pipe.*

end of the tape is an electronic probe. The tape is lowered into the well, and when the probe touches the surface of the water, a light or a bleeper switches on at the top of the well. The operator then holds the tape against the top of the well and reads the measurement. This gives the depth of the water below the top of the well.

The water level must be measured in the pumped well and all observation wells before the start of the test. During the test, the change in water level must be measured regularly in every well. The change will be different in each well.

At the beginning of the test, the water level drops very quickly. Later, it drops more slowly. For this reason, the measurements are taken at short-time intervals early in the test, with the time between measurements increasing as the test progresses. For example, the measurements may be taken as follows:

  (i) Every minute for the first 10 minutes of pumping.

 (ii) Then every 5 minutes until the completion of one hour.

(iii) Then every 15 minutes until the completion of 4 hours.

(iv) Then every 30 minutes until the completion of 8 hours.

 (v) Then every hour for the rest of the test.

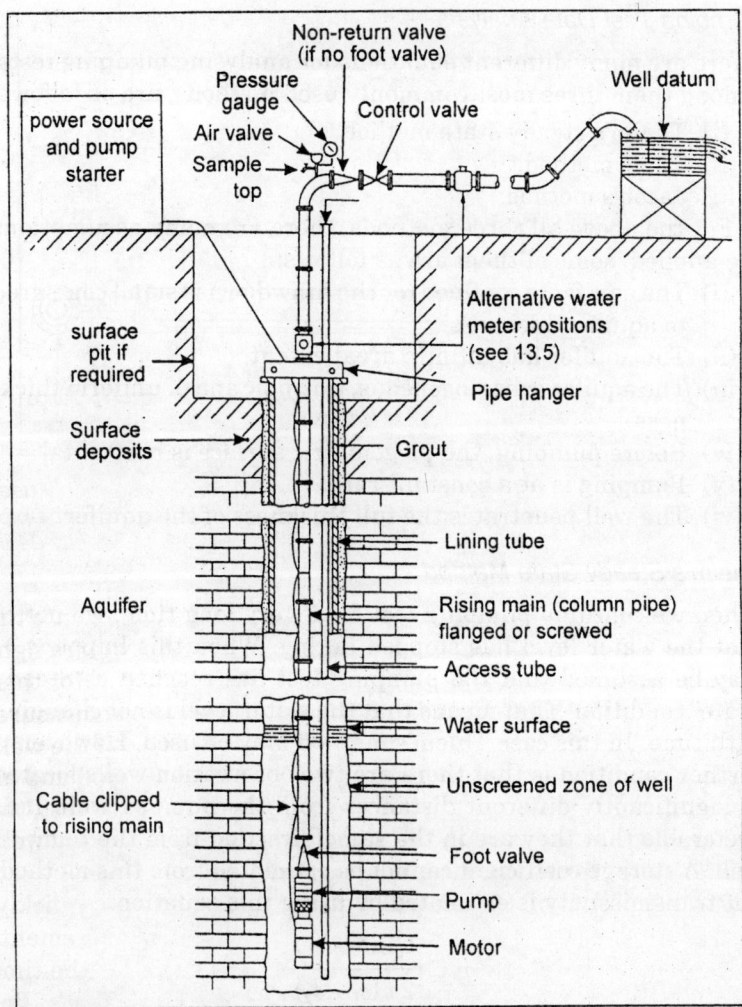

Fig. 37.27: *Typical arrangement of equipment in an unscreened test well.*

It is most important to make sure that an accurate clock is required to make sure the measurements are taken at the correct times.

Pumping Test Measurement

This may be measured with: (a) Weir tank, (b) a flow meter, (c) an orifice weir.

Pumping Test Data Analysis

There are many different equations for analysing pumping tests. Among them three most commonly used methods, are as follows:

  (i) Theim's steady state method.
 (ii) Theis method.
(iii) Jacob's method.

For the above all three methods, there are many assumptions are applied, some of them are as follows:

  (i) The aquifer is confined (or the drawdown is small compared to aquifer thickness).
 (ii) The aquifer has infinite areal extent.
(iii) The aquifer is homogeneous, isotropic and of uniform thickness.
(iv) Before pumping, the piegometric surface is horizontal.
 (v) Pumping is at a constant rate.
(vi) The well penetrates the full thickness of the aquifer.

Theim's Steady State Method

When we continue pumping test for a very long-time, we notice that the water level has stopped falling. When this happens, it may be assumed that the pumping test has reached a "Steady State" condition. That means that the water level is not changing with time. In this case Theim's method may be used. However, a further condition is that there are two observation wells located at significantly different distances from the pumped well. It is preferable that they are in the same direction from the pumped well. A storage coefficient cannot be estimated from this method, but transmissivity is estimated by using this equation.

$$T = \frac{Q\ln (r_2/r_1)}{2\pi (s_1 - s_2)}$$

Where T is transmissivity in m²/d.
Q is the pumping rate in m³/d.
r_1 is the distance of the nearest observation well from the pumped well in metres.
r_2 is the distance of the further observation well from the pumped well in metres.
s_1 is the steady state drawdown at the nearest observation well in metres.

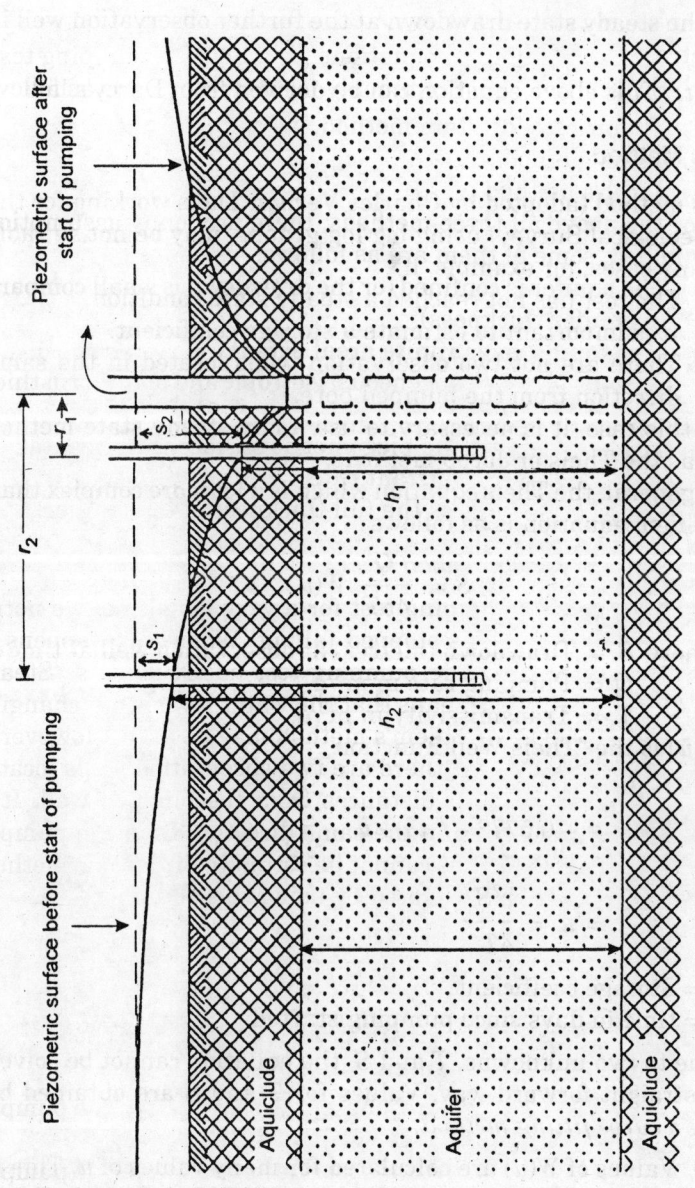

Fig. 37.28: *Cross-section of a pumped confined aquifer.*

s_2 is the steady state drawdown at the further observation well in metres.

Note: This above equation can be derived from Darcy's law.

Theis Method

Theis method (adopted by Charles Vernon Theis working for the US Geological Survey) using log-log graphs, may be not suitable in many cases, for example, if:

(i) The test has not reached a steady state condition.

(ii) It is important to estimate a storage coefficient.

(iii) There are not two observation wells located in the same direction from the pumped bore.

In this case, it is necessary to use a non-steady state method such as the Theis or Jacob methods.

In general, the Theis equation which is a lot more complex than the Thiem equation is as follows

$$S = \frac{Q}{4\pi T} W(u)$$

Where S = Drawdown (metres) at observation well at time t.

Q = Pumping rate in m^3/d.

T = Transmissivity in m^2/d.

$W(u)$ is the "Theis Well Function".

$$= -0.5772 - /nu + u - \frac{u^2}{2.2} + \frac{u^3}{3.3}$$

$$= u = \frac{r^2 S}{4Tt}$$

S = Storage coefficient.

T = time in days since pumping started.

Due to two unknowns, T and S, the equation cannot be solved in a straight forward way. Values for T and S are obtained by using a *curve fitting method*.

(i) Values of $W(u)$ are calculated for many values of u. This is complex, so standard tables are provided as shown in Table 37.1. It is more useful to use $1/u$ instead of u, so these are also given.

Calculate S by Substituting the values of T, r and 1/u in equation $u = \dfrac{r^2 S}{4Tt}$

Table 37.1: Values of the Theis well function $W(u)$ for confined aquifers (after Walton, 1962)

1/u = (n)	u (N)	W(u)	n(1) N(-1)	n(2) N(-2)	n(3) N(-3)	n(4) N(-4)	n(5) N(-5)	n(6) (N-6)	n(7) (N-7)	n(8) N(-8)	n(9) N(-9)	n(10) N(-10)
1.000	1.0	$W(u)=$ 2.194 (-1)	1.823	4.038	6.332	8.633	1.094 (1)	1.324 (1)	1.544 (1)	1.784 (1)	2.015 (1)	2.245 (1)
0.833	1.2	1.584 (-1)	1.660	3.858	6.149	8.451	1.075 (1)	1.306 (1)	1.536 (1)	1.766 (1)	1.996 (1)	2.227 (1)
0.666	1.5	1.000 (-1)	1.465	3.637	5.927	8.228	1.053 (1)	1.283 (1)	1.514 (1)	1.744 (1)	1.974 (1)	2.204 (1)
0.500	2.0	4.890 (-2)	1.223	3.355	5.639	7.940	1.024 (1)	1.255 (1)	1.485 (1)	1.715 (1)	1.945 (1)	2.176 (1)
0.400	2.5	2.491 (-2)	1.044	3.137	5.417	7.717	1.002 (1)	1.232 (1)	1.462 (1)	1.693 (1)	1.923 (1)	2.153 (1)
0.333	3.0	1.305 (-2)	9.057 (-1)	2.959	5.235	7.535	9.837	1.214 (1)	1.444 (1)	1.674 (1)	1.905 (1)	2.135 (1)
0.286	3.5	6.970 (-3)	7.942 (-1)	2.810	5.081	7.381	9.683	1.199 (1)	1.429 (1)	1.659 (1)	1.889 (1)	2.120 (1)
0.250	4.0	3.779 (-3)	7.024 (-1)	2.681	4.948	7.247	9.550	1.185 (1)	1.415 (1)	1.646 (1)	1.876 (1)	2.106 (1)
0.222	4.5	2.073 (-3)	6.253 (-1)	2.568	4.831	7.130	9.432	1.173 (1)	1.404 (1)	1.634 (1)	1.864 (1)	2.094 (1)
0.200	5.0	1.148 (-3)	5.598 (-1)	2.468	4.726	7.024	9.326	1.163 (1)	1.393 (1)	1.623 (1)	1.854 (1)	2.084 (1)
0.166	6.0	3.601 (-4)	4.544 (-1)	2.295	4.545	6.842	9.144	1.145 (1)	1.375 (1)	1.605 (1)	1.835 (1)	2.066 (1)

Table 37.1: Values of the Theis well function $W(u)$ for confined aquifers (after Walton, 1962) *(Contd.)*

$1/u = n$	n	$n(1)$	$n(2)$	$n(3)$	$n(4)$	$n(5)$	$n(6)$	$n(7)$	$n(8)$	$n(9)$	$n(10)$
n	N	$u = N$	$N(-1)$	$N(-2)$	$N(-3)$	$N(-4)$	$N(-5)$	$(N-6)$	$(N-7)$	$N(-8)$	$N(-9)$	$N(-10)$
0.142	7.0	1.155 (–4)	3.738 (–1)	2.151	4.392	6.688	8.990	1.129 (1)	1.360 (1)	1.590 (1)	1.820 (1)	2.050 (1)
0.125	8.0	3.767 (–5)	3.106 (–1)	2.027	4.259	6.555	8.856	1.116 (1)	1.346 (1)	1.576 (1)	1.807 (1)	2.037 (1)
0.111	9.0	1.24 5(–5)	2.602 (–1)	1.919	4.142	6.437	8.739	1.104 (1)	1.334 (1)	1.565 (1)	1.795 (1)	2.025 (1)

How to use table

$N(-2) = N \times 10^{-2}$

For example: What is $W(u)$, if $u = 0.00045$?

$\Rightarrow 0.00045 = 4.5 \times 10^{-4} = 4.5 \, (-4)$ if $N = 4.5$

Look along sow for $N = 4.5$, and down column for $N(-4)$

then, at the intersection, $W(u) = 7.130$

(ii) Plot a graph of $W(u)$ against $1/u$ a log-log paper. This forms a distinctive curve of decreasing gradient as shown in Fig. 37.29.

(iii) Plot drawdown, S (in metres) against time, t (in minutes) also on log-log paper of the same scale. This forms a curve of a very similar shape.

(iv) This is where the curve fitting comes in. Lay the time drawdown curve over the Thies curve. With the axes of the two graphs parallel, adjust the top graph to a position so that most of the data points fall on the Theis curve.

(v) Select a match point. This does not need to be on the curve but can be any point where the graphs overlap. It is better to select a point where $W(u)$ and $1/u$ are equal to either 1 or 10.

(vi) For the same match point determine S and t. Convert the time units for minutes to days by dividing by 1440.

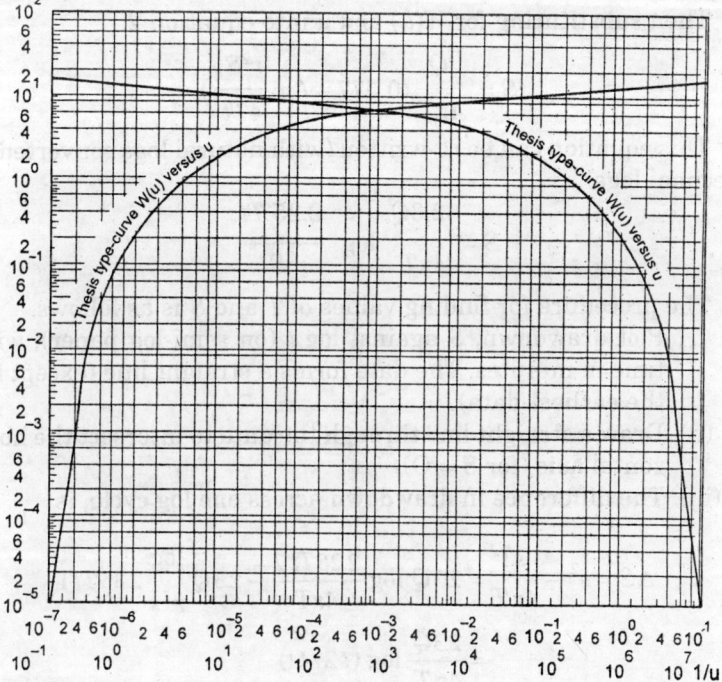

Fig. 37.29: *Theis type curve for W(u) versus u and W(u) versus 1/u.*

(vii) Substitute the value of $W(u)$, S and Q into equation 2 and solve for T.
(viii) Calculate S by substituting the values to T, r and $1/u$ in equation

$$u = \frac{rS}{4Tt}$$

Jacob's Method

Jacob's method using semi-log graphs is an alternate non-steady state method which is based on a simplification of the Theis formula.

After a few minutes into the test, if u is less than 0.01, the terms in 3rd, 4th and later terms in equation

$$W(u) = -0.577 - /u + u - \frac{u^2}{2.21} + \frac{u^3}{3.31} - ....$$

become very small and $W(u)$ is approximately equal to $(-0.577 - \text{Inu})$.

Then substituting for $W(u)$ and μ into equation 2.

$$S = \frac{Q}{4\pi T}(0.577 - /n\frac{r^2S}{4Tt})$$

This equation can be re-written (with natural logs converted to decimal logs) as:

$$S = \frac{2.3Q}{4\pi T} \log \frac{2.25\,Tt}{r^2S}$$

The procedure for finding values of T and S is as follows.
 (i) Plot drawdown, S against $\log t$ (on semi-log paper), with time in minutes. The data forms a straight line (except for the earliest data).
 (ii) Draw a straight line through the data to intersect the horizontal axis (for $S = O$).
 (iii) The difference in drawdown across one log cycle, is

$$\Delta S = s^2 - \frac{s1}{4\pi T} = 2.3Q \log \frac{2.25Tt2}{4\pi T} - 2.3Q \frac{\log}{r^rs} 2.35Tt1$$

$$= \frac{2.3Q}{4\pi T} \log (t2/t1)$$

Rectangular Weir Tank V-Notch

Rectangular weir tanks with discharge capacity of 150 litres/second constructed from steel plate and fitted with a V-notch weir and water level monitor. To prevent surging in the V-notch channel devices should also be present at water well drilling site.

The height of the V-notch are generally two and a half times the streams head, and the upstream head is as the head above the invert of the "V" beyond the influence of the notch. The steel plate at the V-edge is between 1 to 2 millimeters thick, any excess being levelled on the downstream face at an angle of not more than 30 degress to the face of the plate. The weir tank is generally fitted with three baffle plates and water level gauge. The width of the weir box is five times the upstream head.

Over one log cycle of time, $t_2 = 10 \times t_1$, and $\log(t_2/t_1)$
$$= \log 10 = 1.$$

Then
$$\Delta S = \frac{2.3Q}{4\pi T}$$

Therefore, determine ΔS as the difference in drawdown across one log cycle. Then, rearranging above equation to calculate T.

$$T = \frac{2.3Q}{4\pi T \Delta S}$$

(i) The time at which the straight line crosses the horizontal axis is to, determine t_0 and convert the time units from minutes into days by dividing by 1440.

(ii) From equation 6, it can be shown that for

$$S = 0, \frac{2.25 \, Tt_0}{r^2 S} = 1$$

Then the storage coefficient is calculated from:

$$S = \frac{2.25 \, Tt_0}{r^2}$$

38 | Well Development

Well development usually done after the completion to a maximun yield of water, clean and free of suspended materials. An important factor in any well development method is that it should be started slowly and gently and increased in vigour as the well is developed. Well development requires the application of sufficient energy to disturb the natural formation or filter pack so as to free the fines and allow them to be drawn into the well, and to cause the coarser fractions to settle around and stablish the screen. This is usually accomplished by the surging of water into and out of the well and the formation or hydraulic jetting, which depends upon a high velocity water jet discharging through the screen.

As a general rule the amount of development can be estimated from sand sample caught in an Imhoff cone or by other methods on each resumption of pumping. On initiation of interrupted pumping, samples should be caught as frequently as possible as soon discharge starts at each new rate of pumping. Sampling of this type at sand content to occur and will serve as a guide to subsequent sampling and development.

Shortly after the period in which maximum sand content occurs in the discharge for each new rate of pumping, the discharge will become practically sand free until the well is again surged. As development continues, the amount of sand at the maximum will decrease and the time in each discharge interval until water of low sand content is discharge becomes shorter.

The casing and screen diameters, length of screen and character of the formation are determining factors in the selection of application methods of well development.

Fig. 38.1: *V-notch to measure the dischare capacity of borehole (well) in litres/second.*

3″ dia × 1″ dia

Pipe size = 3″ dia; Orifice = 1″ dia

$Q = 8.02 \, KA \, \sqrt{H}$ gallons/min (1 gallon = 3.785 litres)

$Q = 8.02 \, KA \, \sqrt{H} \times 0.063$ litres/sec.

$Q = 0.22 \, \sqrt{H}$ litres/sec.

Height (inch)	Litres/sec	Height (inch)	Litres/sec	Height (inch)	Litres/sec
5	0.491	20	0.984	35	1.300
6	0.539	21	1.000	36	1.320
7	0.582	22	1.030	37	1.339
8	0.620	23	1.060	38	1.356
9	0.660	24	1.078	39	1.374
10	0.696	25	1.100	40	1.390
11	0.730	26	1.122	41	1.409
12	0.760	27	1.143	42	1.426
13	0.793	28	1.164	43	1.443
14	0.823	29	1.185	44	1.460
15	0.852	30	1.205	45	1.476
16	0.880	31	1.225	46	1.492
17	0.908	32	1.245	47	1.500
18	0.933	33	1.264	48	1.524
19	0.959	34	1.283	49	1.540
				50	1.555

3″ dia × 1.1/4″ dia

Pipe size = 3″ dia; Orifice = 1.1/4″ dia

$Q = 8.02 \text{ KA } \sqrt{H}$ gallons/min (1 gallon = 3.785 litres)

$Q = 8.02 \text{ KA } \sqrt{H} \times 0.063$ litres/sec.

$Q = 0.353 \sqrt{H}$ litres/sec

Height (inch)	Litres / sec	Height (inch)	Litres / sec
5	0.789	28	1.870
6	0.865	29	1.900
7	0.934	30	1.933
8	0.999	31	1.965
9	1.059	32	1.996
10	1.116	33	2.000
11	1.170	34	2.058
12	1.222	35	2.088
13	1.273	36	2.118
14	1.321	37	2.147
15	1.367	38	2.176
16	1.412	39	2.204
17	1.455	40	2.232
18	1.498	41	2.260
19	1.539	42	2.287
20	1.579	43	2.315
21	1.618	44	2.342
22	1.656	45	2.368
23	1.693	46	2.394
24	1.729	47	2.420
25	1.765	48	2.446
26	1.800	49	2.471
27	1.834	50	2.496

3″ dia × 1.1/2″ dia

Pipe size = 3″ dia; Orifice = 1.1/2″ dia

$Q = 8.02 \ KA \ \sqrt{H}$ gallons/min (1 gallon = 3.785 litres)

$Q = 8.02 \ KA \ \sqrt{H} \times 0.063$ litres/sec.

$$Q = 0.522 \ \sqrt{H} \ \textbf{litres/sec}$$

Height (inch)	Litres / sec	Height (inch)	Litres / sec
4	1.044	11½	1.770
4 ½	1.107	11³/₄	1.789
5	1.167	12	1.808
5 ½	1.224	12 ¼	1.827
6	1.279	12 ½	1.846
6 ½	1.331	12 ³/₄	1.864
7	1.381	13	1.882
7 ¼	1.406	13 ¼	1.900
7 ½	1.430	13 ½	1.918
7 ³/₄	1.453	13 ³/₄	1.936
8	1.476	14	1.953
8 ¼	1.499	14 ¼	1.971
8 ½	1.522	14 ½	1.988
8 ³/₄	1.544	14 ³/₄	2.004
9	1.566	15	2.022
9 ¼	1.588	15 ¼	2.038
9 ½	1.609	15 ½	2.055
9 ³/₄	1.630	15 ³/₄	2.072
10	1.651	16	2.088
10 ¼	1.671	16 ¼	2.104
10 ½	1.691	16 ½	2.120
10 ³/₄	1.711	16 ³/₄	2.136
11	1.731	17	2.152
11¼	1.751	17.½	2.184

Pipe size = 6″ φ; Orifice size = 3″ φ

$Q = 8.02 \, KA \, \sqrt{H} \times 0.063$ litres/sec.

$Q = 2.09 \, \sqrt{H}$ litres/sec

Height (inch)	Litres / sec	Height (inch)	Litres / sec
8¼	6.0	31½	11.73
9	6.25	33	12.00
9¾	6.52	34¼	12.23
10½	6.77	35¾	12.50
11¼	7.0	37¼	12.76
12	7.24	38¾	13.01
13	7.54	40¼	13.26
13¾	7.75	41¾	13.50
14¾	8.03	43.¼	13.74
15½	8.22	44¾	13.98
16½	8.49	46½	14.25
17½	8.74	48	14.48
18½	8.99	49¾	14.74
19½	9.23	51½	15.00
20½	9.52	53¼	15.25
21¾	9.75	55	15.50
23	10.02	56¾	15.75
24	10.24	58¾	16.07
25¼	10.50	60½	16.25
26½	10.76	62¼	16.49
27¾	11.00	64¼	16.75
29	11.25	66¼	17.01
30¼	11.50	68¼	17.25
		70	17.50

Pipe size = 6″; Orifice = 4″

$Q = 8.02 \text{ KA } \sqrt{H}$

$\quad = 8.02 \times 0.635 \times 12.571 \sqrt{H}$

$\quad\quad = 64.02 \sqrt{H} \quad\quad$ 1 gal = 3.785 litres.

Height	Litres / sec		Height	Litres / sec
6″	9.89		27″	20.98
7″	10.68		28″	21.37
8″	11.42		29″	21.74
9″	12.11		30″	22.12
10″	12.77		31″	22.48
11″	13.39		32″	22.84
12″	13.99		33″	23.19
13″	14.56		34″	23.54
14″	15.11		35″	23.89
15″	15.64		36″	24.23
16″	16.15		37″	24.56
17″	16.65		38″	24.89
18″	17.13		39″	25.22
19″	17.60		40″	25.54
20″	18.06		41″	25.85
21″	18.50		42″	26.17
22″	18.94		43″	26.48
23″	19.36		44″	26.78
24″	19.78		45″	27.09
25″	20.19		46″	27.39
26″	20.59		47″	27.68

Pipe size = 3″; Orifice = 2″

$$Q = 8.02 \ KA \ \sqrt{H}$$
$$= 8.02 \times 0.64 \times 3.142 \ \sqrt{H}$$
$$= 16.127 \ \sqrt{H}$$
$$= 1.016 \ \sqrt{H} \ \text{litres/sec.} \qquad 1 \ \text{gal} = 3.785 \ \text{litres}$$

Height	Litres/sec.	Height	Litres/sec.	Height	Litres/sec.
2″	1.43	21″	4.66	41″	6.5
2¼″	1.52	22″	4.77	42″	6.58
3″	1.76	23″	4.87	43″	6.66
4″	2.03	24″	4.98	44″	6.74
5″	2.27	25″	5.08	45″	6.81
6″	2.5	26″	5.18	46″	6.89
7″	2.68	27″	5.28	47″	6.96
8″	2.87	28″	5.38	48″	7.04
9″	3.0	29″	5.47	49″	7.11
10″	3.2	30″	5.57	50″	7.18
11″	3.36	31″	5.66	51″	7.25
12″	3.52	32″	5.75	52″	7.32
13″	3.66	33″	5.84	53″	7.40
14″	3.8	34″	5.93	54″	7.46
15″	3.93	35″	6.0	55″	7.53
16″	4.0	36″	6.1	56″	7.60
17″	4.2	37″	6.18	57″	7.67
18″	4.3	38″	6.26	58″	7.73
19″	4.43	39″	6.34	59″	7.80
20″	4.54	40″	6.42	60″	7.87

Pipe size = 4″; Orifice = 3″

$$Q = 8.02 \text{ KA } \sqrt{H}$$
$$= 8.02 \times 0.71 \times 7.071 \sqrt{H}$$
$$= 40.263 \sqrt{H} \qquad \text{1 gal = 3.785 litres.}$$

Height	Litres / sec.		Height	Litres / sec.
6″	6.22		29″	13.76
7″	6.72		30″	13.91
8″	7.18		31″	14.14
9″	7.61		32″	14.36
10″	8.03		33″	14.59
11″	8.42		34″	14.81
12″	8.79		35″	15.02
13″	9.15		36″	15.23
14″	9.50		37″	15.44
15″	9.83		38″	15.65
16″	10.15		39″	15.86
17″	10.47		40″	16.06
18″	10.77		41″	16.26
19″	11.07		42″	16.46
20″	11.35		43″	16.65
21″	11.63		44″	16.84
22″	11.91		45″	17.03
23″	12.18		46″	17.22
24″	12.44		47″	17.41
25″	12.69		48″	17.59
26″	12.95		49″	17.77
27″	13.19		50″	17.95
28″	13.44		51″	18.13

Pipe = 4″ φ; Orifice = 2½ φ

$$Q = 8.02 \text{ KA } \sqrt{H}$$
$$= 8.02 \times 0.62 \times 4.91 \sqrt{H} \quad 1 \text{ gal} = 3.785 \text{ litres}$$
$$= 24.414 \sqrt{H}$$
$$= 1.54 \times \sqrt{H}$$

Height	Litres / sec.		Height	Litres / sec.
6″	3.77		25″	7.70
7″	4.07		26″	7.85
8″	4.35		27″	8.00
9″	4.62		28″	8.14
10″	4.87		29″	8.29
11″	5.10		30″	8.43
12″	5.33		31″	8.57
13″	5.55		32″	8.71
14″	5.76		33″	8.84
15″	5.96		34″	8.98
16″	6.16		35″	9.11
17″	6.35		36″	9.24
18″	6.53		37″	9.36
19″	6.71		38″	9.49
20″	6.88		39″	9.61
21″	7.05		40″	9.74
22″	7.22		41″	9.86
23″	7.38		42″	9.98
24″	7.54		43″	10.09

Pipe Size = 6″; Orifice = 4½″

$$Q = 8.02 \text{ KA } \sqrt{H}$$
$$= 8.02 \times 0.71 \times 15.91 \sqrt{H}$$
$$= 90.59 \sqrt{H} \text{ GPM} \qquad 1 \text{ gal} = 3.785 \text{ litres}$$
$$= 5.7 \sqrt{H} \text{ litres/sec.}$$

Height	Litres / sec.		Height	Litres / sec.
5″	12.74		22″	26.73
6″	13.96		23″	27.33
7″	15.08		24″	27.92
8″	16.12		25″	28.50
9″	17.10		26″	29.06
10″	18.02		27″	29.61
11″	18.90		28″	30.16
12″	19.74		29″	30.69
13″	20.55		30″	31.22
14″	21.32		31″	31.73
15″	22.07		32″	32.24
16″	22.80		33″	32.74
17″	23.50		34″	33.23
18″	24.18		35″	33.72
19″	24.84		36″	34.20
20″	25.49		37″	34.67
21″	26.12		38″	35.13

Flow measurement by 90° V-Notch

(1 US GPM = 3.8 Litres/sec.)

Inch	Cm	Gpm	Litres/sec.	Inch	Cm	Gpm	Litres/sec.
1	2.54	2.19	0.1387	7¼	18.415	310	19.63
1¼	3.175	3.83	0.2425	7½	19.05	338	21.41
1½	3.81	6.05	0.383	7¾	19.685	357	23.24
1¾	4.445	8.89	0.563	8	20.32	397	25.14
2	5.080	12.4	0.785	8¼	20.955	429	27.17
2¼	5.715	16.7	1.058	8½	21.590	462	29.26
2½	6.35	21.7	1.374	8¾	22.225	498	31.54
2¾	6.985	27.5	1.742	9	22.86	533	33.756
3	7.620	34.2	2.166	9¼	23.495	571	36.163
3¼	8.255	41.8	2.65	9½	24.13	610	38.63
3½	8.890	50.3	3.19	9¾	24.765	651	41.23
3¾	9.525	59.7	3.78	10	25.4	694	43.95
4	10.160	70.2	4.45	10¼	26.035	739	46.103
4¼	10.745	81.7	5.17	10½	26.67	784	49.653
4½	11.43	94.2	5.97	10¾	27.305	832	52.693
4¾	12.065	108	6.84	11	27.94	880	55.733
5	12.70	123	7.79	11¼	28.575	932	59.03
5¼	13.335	139	8.80	11½	29.21	984	62.32
5½	13.970	156	9.88	11¾	29.85	1039	65.80
5¾	14.605	174	11.02	12	30.48	1094	69.28
6	15.24	193	12.22	12¼	31.115	1153	73
6¼	15.875	214	13.55	12½	31.75	1212	76.8
6½	16.51	236	14.95	12¾	32.385	1274	80.72
6¾	17.145	260	16.47	13	33.02	1337	84.67
7	17.780	284	17.99	13¼	33.655	1469	93.03

90° V-Notch Weir

Flow in litres/second from heads in millimetres

Head (mm)	Litres/sec.	Head (mm)	Litres/sec.	Head (mm)	Litres/sec.	Head (mm)	Litres/sec.
10	0.01	41	0.47	72	1.92	103	4.70
11	0.02	42	0.50	73	1.99	104	4.81
12	0.02	43	0.53	74	2.06	105	4.93
13	0.03	44	0.56	75	2.13	106	5.05
14	0.03	45	0.59	76	2.20	107	5.17
15	0.04	46	0.63	77	2.27	108	5.29
16	0.04	47	0.66	78	2.34	109	5.41
17	0.05	48	0.70	79	2.42	110	5.54
18	0.06	49	0.73	80	2.50	111	5.66
19	0.07	50	0.77	81	2.58	112	5.79
20	0.08	51	0.81	82	2.66	113	5.92
21	0.09	52	0.85	83	2.74	114	6.06
22	0.10	53	0.89	84	2.82	115	6.19
23	0.11	54	0.94	85	2.91	116	6.32
24	0.12	55	0.98	86	2.99	117	6.46
25	0.14	56	1.02	87	3.08	118	6.60
26	0.15	57	1.07	88	3.17	119	6.74
27	0.17	58	1.12	89	3.26	120	6.88
28	0.18	59	1.17	90	3.35	121	7.03
29	0.20	60	1.22	91	3.45	122	7.17
30	0.22	61	1.27	92	3.54	123	7.32
31	0.23	62	1.32	93	3.64	124	7.47
32	0.25	63	1.37	94	3.74	125	7.62
33	0.27	64	1.43	95	3.84	126	7.78
34	0.29	65	1.49	96	3.94	127	7.93
35	0.32	66	1.54	97	4.04	128	8.09
36	0.34	67	1.60	98	4.15	129	8.25
37	0.36	68	1.66	99	4.26	130	8.41
38	0.39	69	1.73	100	4.36	131	8.57
39	0.41	70	1.79	101	4.47	132	8.74
40	0.44	71	1.85	102	4.59	133	8.90

90° V-Notch Weir

Flow in litres/second from heads in millimetres. *(Contd.)*

Head (mm)	Litres / sec.	Head (mm)	Litres / sec.	Head (mm)	Litres / sec.	Head (mm)	Litres / sec.
134	9.07	164	15.03	194	22.88	224	32.77
135	9.24	165	15.26	195	23.17	225	33.14
136	9.41	166	15.49	196	23.47	226	33.51
137	9.59	167	15.73	197	23.77	227	33.88
138	9.76	168	15.96	198	24.07	228	34.25
139	9.94	169	16.20	199	24.38	229	34.63
140	10.12	170	16.44	200	24.69	230	35.01
141	10.30	171	16.69	201	25.00	231	35.39
142	10.49	172	16.93	202	25.31	232	35.78
143	10.67	173	17.18	203	25.62	233	36.16
144	10.86	174	17.43	204	25.94	234	36.55
145	11.05	175	17.68	205	26.26	235	36.94
146	11.24	176	17.93	206	26.58	236	37.34
147	11.43	177	18.19	207	26.90	237	37.74
148	11.63	178	18.45	208	27.23	238	38.13
149	11.83	179	18.71	209	27.56	239	38.54
150	12.03	180	18.97	210	27.89	240	38.94
151	12.23	181	19.23	211	28.22	241	39.35
152	12.43	182	19.50	212	28.56	242	39.76
153	12.64	183	19.77	213	28.90	243	40.17
154	12.84	184	20.04	214	29.24	244	40.58
155	13.05	185	20.31	215	29.58	245	41.00
156	13.26	186	20.59	216	29.92	246	41.42
157	13.48	187	20.87	217	30.27	247	41.84
158	13.69	188	21.15	218	30.62	248	42.27
159	13.91	189	21.43	219	30.97	249	42.70
160	14.13	190	21.72	220	31.33	250	43.13
161	14.35	191	22.00	221	31.69	251	43.56
162	14.58	192	22.29	222	32.05	252	43.99
163	14.80	193	22.58	223	32.41	253	44.43

90° V-Notch Weir

Flow in litres/second from heads in millimetres. *(Contd.)*

Head (mm)	Litres/sec.	Head (mm)	Litres/sec.	Head (mm)	Litres/sec.	Head (mm)	Litres/sec.
254	44.87	270	52.27	286	60.37	302	69.17
255	45.31	271	52.76	287	60.90	303	69.74
256	45.76	272	53.25	288	61.43	304	70.32
257	46.21	273	53.74	289	61.96	305	70.90
258	46.66	274	54.23	290	62.50	306	71.48
259	47.11	275	54.73	291	63.04	307	72.07
260	47.57	276	55.23	292	63.58	308	72.65
261	48.03	277	55.73	293	64.13	309	73.24
262	48.49	278	56.23	294	64.68	310	73.84
263	48.95	279	56.74	295	65.23	311	74.44
264	49.42	280	57.25	296	65.78	312	75.04
265	49.89	281	57.76	297	66.34	313	75.64
266	50.36	282	58.28	298	66.90	314	76.24
267	50.83	283	58.80	299	67.46	315	76.85
268	51.31	284	59.32	300	68.03		
269	51.79	285	59.84	301	68.60		

Fig. 38.2: *Air lift well development in Oman.*

1. Interrupted over pumping
2. Surging and bailing utilising bailer.
3. Surging and bailing utilising surge block.
4. Mechanical surging and pumping method.
5. Hydraulic jetting method.
6. Single pipe system open to the atmosphere.
7. Single pipe system close to the atmosphere.
8. Two pipe system.
9. Washing with water.
10. Washing with chemicals, etc.

Interrupted Overpumping

The development process may include development by interrupted pumping. The pumping shall be done at rates up to 2 times the design capacity. The pumping should be carried at out in at least 5 steps, which should include pumping rates of 0.25, 0.5, 1, 1.5 and 2 times the design capacity, with no check valve nor foot valve present. Pumping shall be conducted in 5 minutes cycles, and shall continue for a minimum of 2 hours but as per applicable.

Surging and Bailing (Utilising Bailer)

The development process may include surging and bailing the well. The surging shall be accomplished by utilising the bailer as a suging device. If fine sediments have been drawn into the well and have settled on the bottom and accumulated to a depth where they block 10 percent of more of the total screen length, the well may be bailed or otherwise cleaned to the bottom before resumption of surging. On completion of development the well shall be cleaned to the bottom.

Surging and Bailing (Utilising Surge Block)

The development process may be carried out by surging and bailing the well. The surging shall be done by single or double solid (or valved) surge block, surging shall short at the bottom of the lowest screen in the well and proceed upwards.

Mechanical Surging and Pumping Method

The development process may include surging and pumping the well. The surging shall be done by either a solid or valved surge block. The pumping shall be done through the surge block which incorporates a piece of the suction pipe in the fabrication of the block. Pumping shall be done simultaneously with the surging at rates up to half of the design capacity. Fine sediments drawn into the well shall be pumping out periodically before such accumulation reached 10% of the screen length. Upon completion of the development work the well shall be cleaned to the bottom.

Hydraulic Jetting Method

Development may be accomplished by simultaneous high-velocity, horizontal-jetting and pumping. The outside diameter of the jetting tool shall be one inch less in diameter than the screen inside diameter. The maximum exit velocity of the jetting fluid at the jet nozzle shall be 45 m/sec (150 ft). The tool shall be rotated at a speed less than 1 rpm. It shall be positioned at one level for not less than two minutes and then shall be moved to the next level which shall be no more than 15 cm (6 inches) vertically from the preceding jetting level. The jetting shall proceed from the bottom of the screen to the top. Pumping from the well shall be at a rate of 5 to 15 percent more than the rate at which water is introduced through the jetting tool. Water to be used for jetting must contain less than 100 ppm suspended solids.

Single Pipe System Open to the Atmosphere

Development may be done by utilisation of a single pipe air pumping system using the casing or the borehole itself as the eductor line. The compressors, air lines, hoses, fitting, etc. shall be of adequate size to pump the well by the air lift principle at 1.5 to 2 times the design capacity to the well. Pumping the well continue with air until the well is developed to the point that it yields clear, sand-free water. He shall than shutoff the air and allow water in the well to return to a static condition. He shall than re-open the valve and reintroduce air into the well until water is again brought to the surface by the air lift, at which time he will close the air valve and allow the water to drop back down the well and return to a static condition. He shall repeat this lifting and dropping of the coloum of water until the water in the well becomes turbid at which time he will continuously pump the well with air until it again yields clear sand-free water. We shall repeat the above operations until the well no longer produces fine material. When it is surged and backwashed as described above.

The bottom of the air line may be placed at different levels to facilitate development of all intake areas and multiple water producing zones, and the process repeated until all zones yield water free of turbidity when surged and backwashed.

Single Pipe System Close to Atmosphere

We may install a suitable valve on the discharge line leading from the top of casing, secure the air line into blowing "T" or "L" affixed to a valved air connection on the top of the casing. We shall then close the valve on the discharge line, attach the air hose to the valved fitting and introduce air into the well, forcing down the column of water in the well. Care shall be exercised to prevent air form entering the water-bearing formation. This shall be accomplished by the installation of a separate pipe, open to the atmosphere at the top, and installed in the well to a point ten feet above the water-bearing zone. When the water-level in the well is forced down to the bottom of this air release pipe, the discharge valve shall be opened and the water allowed to rise back to the static level. This procedure may be repeated and/or alternated with the *single pipe system open to the atmosphere* technique. A pressure gauge and relief. Valve shall be installed at the top of the casing when this system is used.

Two Pipe System

The development process may be carried out by the utilisation of an air injection pipe and an air and water eductor line. The compressors, air lines, hoses, fitting, etc. shall be of adequate size to pump the well by the *air-lift method* at 1.5 to 2 times the design capacity of the well. We initially develop the well as outlined above in the *single pipe system closed to the atmosphere*, with the air line introducing air into the eductor line at a point above the bottom of the eductor line. When the well yields clear sand-free water, the air line shall be lowered to a point below the bottom of the educator line and the air introduced until the water between the eductor pipe and the casing is raised to the surface. At this time the air line shall be raised back up into the eductor line causing the water to be pumped from the well through the eductor line. The procedure of alternating the relative positions of the air and eductor line shall be repeated until the water yielded by the remains clear when the well is surged and backwashed by this technique.

Washing with Water

Clean, clear water may be circulated to remove sediment from the well. A pump of sufficient size shall be utilised for the washing process which will agitate the formation for the purpose of preventing bridging of the sand particles and removing a large portion of the finer material.

Washing with Chemicals

Where applicable and required mud dispersing agent (such as glassy phosphate), acids for washing limestone and other chemicals applicable to standard procedures may be used.

DISINFECTION PROCEDURE

General

In general, disinfection will normally only be required in production type wells.

Chlorine or other compounds shall be used as disinfectants general aspects of well disinfection are:

(a) Inspection for and removal of foreign matter is a necessary per-requisite to well disinfection.

(b) The specific time in the well construction schedule at which the disinfection is required.

(c) The amount of disinfecting agent to be used and the contact time required.

(d) The establishment of equipment and work procedures required in carrying out the disinfection.

Well cleaning is a necessary part of well construction. Contaminants in the form of grease, oil, soil, and other foreign substances can harbour and protect bacteria from subsequent disinfection.

Delivery and Storage

The disinfectant usually delivered to the site of the work in original closed containers bearing the original label indicating the percentage of available chlorine. Chlorine compounds in dry form shall not be stored for more than one year and storage of liquid compounds shall not exceed 60 days. During storage, disinfectants shall not be exposed to the atmosphere or to direct sunlight.

Where disinfection with a dry chlorine compound is necessary or permitted without first preparing a liquid solution, appropriate means must be provided to achieve a relatively even application of the compound to the bottom of the well screen and throughout the well. Pellets or powdered chlorine compound must be used with a mechanical carrier.

Mixing of Solutions

Usually, a solution is prepared prior to placement of the disinfectant into the well and unless a solution is evenly distributed the disinfectant may more latterly rather than to the bottom of the well. An appropriate tremmie device, hose or pipe, should be employed to ensure proper distribution of the disinfectant. Agitation through use of a bailer, surge block or by intermittent stopping and starting of a test pump, is recommended to force some of the solution into the water-bearing formation around the well. However, agitation is not an alternate to adequate distribution of the concentrated chlorine solution.

Unless suspended by governmental regulation, the quantity of chlorine compounds used for disinfection shall be sufficient to produce a minimum of 50 ppm available chlorine in solution when mixed with the total volume of water in the well. A 50 ppm solution should result from utilising quantities of chlorine compounds, proportional the depth of water.

Placement in the Well

If a test pump is available during disinfection it provides a convenient means for application of the disinfecting solution to the dry part of the well. The discharge piping of either a test pump or permanent pump should incorporate a tap and hose connection on the pump side of a valve in the discharge piping, to facilitate application of the chlorine solution in the well to the dry part of the well through a hose. Intermittent pump operation for surging will not interfere with hose application of the solution to dry parts of the well, and in keeping such parts wet for an adequate period of time. When a pumped supply of dilute chlorine solution is not available for this purpose, a separate tank and gravity system will be needed.

Disinfection of Flowing Artesian Wells

The disinfection of flowing artesian wells with dry chlorine compounds should be done using a doubly capped, perforated pipe container filled with a granular chlorine compound. It shall be placed at a point on or below the top of the producing horizon. This process shall be repeated as often as necessary to achieve and maintain the standard 50 ppm concentration for a period of not less than one hour.

If the flow from the well is able to be controlled by capping then a stock chlorine solution shall be injected, under pressure, by means of a drop pipe to the bottom of the well. The cap shall be equipped with a suitable one inch valve. After the injections is complete air shall be injected for agitation while simultaneously opening the valve in the cop permitting the chlorine solution to be dispersed to the surface. The valve shall then be closed and the flow stopped. The chlorine connected shall be maintained at 50 ppm for six hours. In the event flow can be controlled by a suitable the chlorine treatment can be conducted as though the well was non-flowing.

39 | Tube Well Maintenance

Maintenance of tube wells must be done to run them properly in simple way.

1. The cleaning of the screens is done by the method known as *surging*. In this method the water is forced inside the well and taken out alternatively at very high rate. Sometimes compressed air is also passed inside the tube well before pumping water iniside it. This method removes the clay particles from the screens.

2. Solid carbon dioxide (dry ice) is dropped inside the tube well and it is closed from top by lightening cap. The dry ice vapourises immediately and increases the pressure inside the tube well sufficient to force away the soil particles from the screens.

3. The rusted screens are cleaned with the help of sulphuric acid (H_2SO_4) or hydrochloric acid (HCl), but totally rusted parts should be replaced with new parts.

Hydrogeology: Objective Style

1. Hill method is used for computing

   (a) Safe yield   (b) Specific yield
   (c) Specific retentions   (d) Precipitation

2. The amount of water which can be withdrawn from a ground water basic annually without producing an undesired result is known as

   (a) Specific yield   (b) Base yield
   (c) Safe yield   (d) Specific storage

3. Permanent depletion of safe yield is often referred to as

   (a) Mining   (b) Winning
   (c) Drawdown   (d) Hydraulic gradient

4. When sodium adsorption ratio (SAR) is more than 26th water class will be

   (a) Good   (b) Excellent
   (c) Fair   (d) Poor

5. Irrigation water having less than 20 percent sodium is classified as

   (a) Good   (b) Excellent
   (c) Permissible   (d) Doubtful

6. In the chemical characteristics of drinking water standards the upper limit of arsenic constituents is

   (a) 1 ppm   (b) 1.5 ppm
   (c) 0.05 ppm   (d) 3.0 ppm

7. In the drinking water standard, the upper limit of turbidity is .... (silica scale)

   (a) 20 ppm   (b) 10 ppm

   (c) not objectionable   (d) 1 ppm

8. Ground water having 32 percent of sodium is said to be

   (a) Good   (b) Unsuitable

   (c) Permissible   (d) Doubtful

9. High salinities are found in soils and ground water of

   (a) Tropical climates   (b) Temperate climates

   (c) Arid climates   (d) Sub-tropical climates

10. Which ions are more commonly added to ground water by solution in limestone terrenes

   (a) K and CO_3 (b) Ca and CO_3

   (c) K and NO_3 (d) Mg and Cl

11. Measures of water quality includes .... analyses

   (a) Chemical   (b) Physical

   (c) Sanitary and biological   (d) All of the above

12. Which one is most common chemical constituents of ground water

   (a) $CaCO_3$ (b) $Mg\,CO_3$ (c) $NaSO_4$ (d) KCl

13. Total hardness of the water can be expressed as

   (a) $TH = Ca \times \dfrac{CaCO_3}{Ca} + Mg \times \dfrac{CaCO_3}{Mg}$

   (b) $TH = Mg \times Ca\, \dfrac{CaCO_3}{Ca + Mg}$

   (c) $TH = \dfrac{MgCO_3 + CaCO_3}{Ca \times Mg}$

   (d) all of the above

14. In the physical analysis of ground water turbidity is a measure of the

   (a) Velocity of water

   (b) Suspended and colloidal matter in water

   (c) Organic matter in water

   (d) *b* and *c* both

15. Ground water level fluctuations are due to
    - (a) Evapotranspiration
    - (b) Atmospheric pressure
    - (c) Tides and earthquakes
    - (d) All of the above

16. Magnitude of transpiration fluctuations of water table depends upon the
    - (a) Type of vegetation
    - (b) Season
    - (c) Weather
    - (d) All of the above

17. Portion of stream flow coming from ground water is known as
    - (a) Ground flow
    - (b) Base flow
    - (c) River flow
    - (d) Flood flow

18. The water stored and released after the flood is referred to as
    - (a) Storage coefficient
    - (b) Specific yield
    - (c) Specific retention
    - (d) Bank storage

19. A stream which is receiving ground water discharge is known as
    - (a) Effluent stream
    - (b) Influent stream
    - (c) Lost stream
    - (d) Bank storage

20. A variation of the hydraulic rotary method is known as
    - (a) Cable tool method
    - (b) Hydraulic rotary method
    - (c) Reverse rotary method
    - (d) *a* and *b* both

21. A well whose length of water entry is less than the aquifer which it is penetrates is known as
    - (a) Penetrating well
    - (b) Partially penetrating well
    - (c) Multiple well
    - (d) Jetted well

22. At a given point the distance from which water level lowers is known as
    - (a) Drawdown
    - (b) Hydraulic gradient
    - (c) Lowering water table
    - (d) Permeability gradient

23. The lower the degree of saturation the .... the permeability
    - (a) Lower
    - (b) Higher
    - (c) Constant
    - (d) *a* and *b* both

24. Which one is correct statement

    (a) Permeabilities associated with unsaturated flow are more than those for saturated flow.
    (b) Permeabilities associated with unsaturated flow are less than those for saturated flow.
    (c) Permeabilities of saturated and unsaturated flow are same.
    (d) None of the above.

25. Experimental evidence indicates accelerated flow

    (a) Obeys Darcy's law but with a different permeabilities that for saturated flow.
    (b) Obeys Darcy's law with the same permeability for saturated flow.
    (c) Does not obey Darcy's law.
    (d) None of the above.

26. Dating of ground water is possible by using

    (a) Artificial tritium
    (b) Natural tritium as a tracer
    (c) Copper sulphate solution
    (d) Cobalt as a tracer

27. The permea-metre used to measure permeabilities of consolidated or unconsolidated formations under law heads is

    (a) Constant head permea-meter
    (b) Falling head permea-meter
    (c) Non-discharging permea-meter
    (d) Consolidated or unconsolidated permea-meter

28. Viscosity of the fluid .... with temperature
    (a) Does not vary   (b) Varies inversely
    (c) Varies proportionally (d) None of the above

29. The coefficient of transmissibility (T) equals the

    (a) $T = \dfrac{b}{k}$ (b) $T = QA$

(c) $T = Kb$ (d) None of the above

30. Relation between field and laboratory coefficient of permeabilities can be expressed without reference to viscosity table as

    (a) $\dfrac{K_s}{K_f} = \dfrac{M_{60}}{M_f}$ (b) $\dfrac{K_s}{K_f} = \dfrac{M_f}{M_{20}}$

    (c) $\dfrac{K_s}{K_f} = \dfrac{M_f}{M_{60}}$ (d) $\dfrac{K_f}{K_s} = \dfrac{M_{60}}{M_f}$

31. Darcy's law is applicable in all natural ground water motion when
    (a) $N_R < 1$ (b) $N_R > 1$
    (c) $N_R = 1$ (d) $N_R = 1$ to 1000

32. .... serves as a criterion to distinguish between laminar and turbulent flow
    (a) Coefficient of permeability
    (b) Coefficient of transmissibility
    (c) Poiseuille's number
    (d) Reynold number

33. Reynold number can be expressed as

    (a) $f = \dfrac{d\,\Delta P}{2\,plV^2}$ (b) $NR = Q R\,\dfrac{1}{L}$

    (c) $NR = \dfrac{PvD}{M}$ (d) $NR = \dfrac{M}{PvD}$

34. Darcy's law is only applicable for
    (a) Turbulent flow   (b) Laminar flow
    (c) Turbidite flow   (d) None of the above

35. Dracy's law when written with a negative sign, as $V = -K(dh/dl)$ indicates that the flow is in direction of
    (a) increasing head   (b) decreasing head
    (c) constant head   (d) none of the above

36. An imaginary surface in confined aquifer coinciding with the hydrostatic pressure level of the water in the aquifer is known as

(a) Confining surface   (b) Water table
(c) Piezometric surface (d) Hydrostatic surface

37. "Recharge area" is that region which

   (a) Supplies water to a confined aquifer
   (b) Supplies water to unconfined aquifer
   (c) Supplies water to perched aquifer
   (d) Takes water from confined aquifer

38. Where ground water is confined under pressure greater than atmospheric by overlying, relatively impermeable strata such aquifer are known as

   (a) Confined aquifers   (b) Artesians
   (c) Pressure aquifers   (d) All of the above

39. Porosity and yield of sandstone and conglomerate have been reduced by the

   (a) Difference in grain size (b) Sorting of grains
   (c) Presence of cement   (d) All of the above

40. Large springs are frequently found in

   (a) Limestone area   (b) Sandstone area
   (c) Plain area   (d) All of the above

41. The term "Lost river" has been applied to a stream

   (a) Which disappears completely underground in igneous terrain.
   (b) Which flows within the dead valley.
   (c) Which disappears completely underground in a limestone terraine.
   (d) All of the above.

42. Which will be correct sequence in order of increasing specific yield

   (a) Sand-gravel-clay   (b) Silt-gravel-clay
   (c) Clay-sand-gravel   (d) None of the above

43. Tick the correct sequence of the sediments in order of decreasing porosity

   (a) Clay-silt-gravel   (b) Gravel-uniform sand-clay
   (c) Limestone-clay-soil   (d) Gravel-caly-shale

44. Which is correct statement about saturated zone

   (a) The zone of saturation consists of interstices occupied partially by water and by air.

(b) Water occurring in the zone of saturation is known as capillary water.

(c) In the zone of saturation all of this water may not removed from the ground by drainage or pumping from a well due to molecular and surface lension forces.

(d) All of the above.

45. Porosity is a direct measure of the water contained per unit volume in

(a) Saturated zone   (b) Unsaturated zone

(c) Aerated zone   (d) Both *b* and *c*

46. Specific retention can be expressed as

(a) $Sr = \dfrac{Y}{100\,W}$ (b) $Sr = \dfrac{100\,W}{Y}$

(c) $Sr = \dfrac{V}{100\,W}$ (d) $Sr = \dfrac{100\,W}{V}$

47. Which of the following shows the highest porosity next to soil

(a) Gravel   (b) Shale

(c) Silt   (d) Uniform sand

48. Which of the following is least porous material

(a) Silt   (b) Clay   (c) Sandstone   (d) Shale

49. Which of the following is considered as vadose water

(a) Capillary water   (b) Pellicular water

(c) Gravitational water   (d) All of the above

50. Which of the following is water of aeration zone

(a) Capillary water   (b) Pellicular water

(c) Gravitational water   (d) All of the above

51. Excess soil water which drains through the soil under the influence of gravity is known as

(a) Soil water   (b) Gravitational water

(c) Ground water   (d) Capillary water

52. Darcy's law can be expressed in general terms as

(a) $Q = KA\,\dfrac{dh}{dl}$ (b) $Q = KA\,\dfrac{LL}{dl}$

(c) $Q = KL = \dfrac{dA}{dt}$ (d) $K = Q \sim \dfrac{1}{L}$

53. Darcy's law states, that the flow rate through porous media is

(a) Inversely proportional to the head and proportional to the length of the flow path.

(b) Proportional to the head and inversely proportional to the length of the flow path.

(c) Directly proportional to the headloss and length of flow path.

(d) All of the above.

54. K is a coefficient of permeability depends on the

(a) Nature of the sand

(b) Thickness of the sand bed

(c) a and b both

(d) Stratigraphic position of the bed

55. Fresh surface water is .... than sea water

(a) Heavier   (b) Lighter

(c) Colder   (d) All of the above

56. Equation of "Jacob method of solution" for pumping test is

(a) $T = \dfrac{264\,Q}{\Delta L}$ (b) $F(v) = \dfrac{L_0 - L}{\Delta}$

(c) $T = \dfrac{114.6\,Q}{L_0 - L}$ (d) $T = \dfrac{\Delta L}{264\,Q}$

57. General equation for steady flow in isotropic media is

(a) $\dfrac{d^2h}{d^2x} + \dfrac{d^2h}{d^2y} + \dfrac{d^2h}{d^2z} = 0$ (b) $V = \dfrac{kdh}{ds}$

(c) $\dfrac{d^2h}{dx^2} + \dfrac{d^2h}{dy^2} + \dfrac{d^2h}{dz^2} = 0$ (d) None of the above

58. In the equation $L_0 - L = \dfrac{114.60}{T}\,W\,(m)$ is

(a) Drawdown   (b) Velocity

(c) Well function   (d) Hydraulic gradient

59. The purpose of a test hole before drilling a well in a new area is

    (a) To determine depths to ground water
    (b) Quality of ground water
    (c) Physical characters and thickness of aquifer
    (d) All of the above

60. Which will be suitable well in unconsolidated aquifer, where water table is at shallow depth

    (a) Dug well   (b) Jetted well
    (c) Bored well   (d) Driven well

61. Steady flow equation is

    (a) $\dfrac{\delta vx}{\delta z} + \dfrac{\delta vy}{\delta x} + \dfrac{\delta vz}{\delta y} = 0$ (b) $\dfrac{\delta h}{\delta h} = 0$

    (c) $\dfrac{\delta vx}{\delta x} + \dfrac{\delta vy}{\delta x} + \dfrac{\delta vz}{\delta y} = 0$ (d) $Q = K\Delta\,\dfrac{dh}{dl}$

62. By using constant head permeametor the measurement of permeability can be obtained by the equation

    (a) $K = \dfrac{VL}{Ath}$ (b) $K = \dfrac{Ath}{VL}$

    (c) $K = \dfrac{Al}{2at}$ in $\dfrac{h_0}{h}$ (d) $Q = K\Delta\,\dfrac{dl}{dl}$

63. A pumping test method has an advantage of avoiding curve and unrestricted in its application is

    (a) Jacob method   (b) Chow method
    (c) Theis method   (d) Darcy method

64. 1 darcy is equal to

    (a) 18.10 ks   (b) 18.18 ks
    (c) 18.2 ks   (d) 25.2 ks

65. A recharge well is that well through which

    (a) Water is added to an aquifer
    (b) Water is lost from an aquifer
    (c) Water supply is impossible
    (d) Water supply is possible

66. Temporary hardness of water is due to the

    (a) Bicarbonates of Cu, Mg   (b) Bicarbonates of Ca, Mg
    (c) Carbonates of Fe, Si   (d) Sulphate

67. Pressure of hardness in water depends upon the

    (a) Presence if Ca and Mg salts
    (b) Absence of the Ca and Mg salts
    (c) Pressure of Mg, Fe salts
    (d) Absence of NaCl

68. Water containing less than 1 gram of salts per kg of water is classed as

    (a) Salt water   (b) Fresh water
    (c) Hot water   (d) Cold water

69. Water having least impurities is

    (a) Atmospheric water   (b) Surface water
    (c) Underground water

70. The fastest method for drilling is

    (a) Cable tool method   (b) Boring method
    (c) Hydraulic rotary method   (d) None of the above

─────────────── **ANSWERS** ───────────────

1. (a)	2. (c)	3. (a)	4. (d)	5. (b)
6. (c)	7. (b)	8. (a)	9. (c)	10. (b)
11. (d)	12. (d)	13. (a)	14. (b)	15. (d)
16. (d)	17. (b)	18. (d)	19. (a)	20 (c)
21. (b)	22. (a)	23. (a)	24 (a)	25. (a)
26. (b)	27. (a)	28. (b)	29. (c)	30. (b)
31. (a)	32. (d)	33. (c)	34. (b)	35. (b)
36. (c)	37. (a)	38. (d)	39. (c)	40. (a)
41. (c)	42. (c)	43. (a)	44. (c)	45. (a)
46. (d)	47. (c)	48. (d)	49. (d)	50. (d)
51. (b)	52. (a)	53. (b)	54. (c)	55. (b)
56. (a)	57. (c)	58. (c)	59. (d)	60. (c)
61. (c)	62. (a)	63. (b)	64. (c)	65. (a)
66. (b)	67. (a)	68. (b)	69. (a)	70. (c)

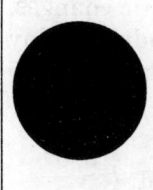

Important Tips

> While planning a township, an adequate source of water supply are found to be lie within the rocks forming the country.

> The engineers build-up dams, barrages and embankments to control the natural flow of running water and they drive tunnels as well as shafts, galleries and draft in mines to create void space within the solid rocks.

> Engineers ventures to protect the sea-shores from the fury of the dashing waves and breakers, and also prevent the accumulation of sediments in the neighbourhood and ports.

> As the geologist analyses the condition of area selected for the site, the engineer will consider as to how to improve the condition and to adjust them, as so to make them suitable, to his plan.

> The geological factor upon which stability of slope depends upon the nature of the rock occurring along the slope or cutting wall. The geological structure of the country-rocks and the prevailing ground water condition along the slope.

> Physical landscape is largely the result of weathering and erosion.

> A formula is given by ranking for the minimum depth of foundation of which bridge peir, buildings, dams, etc. is recommended as:

$$\frac{P}{W} \times \left(\frac{1 - \sin \theta}{1 + \sin \theta} \right)$$

> Bearing capacity of black soils can be improved by draining subsoil water, black soils can undergo volumetric changes. It has a tendency of swelling and shrinking due to clay portion, and safe bearing capacity of black soil varies from 5 to 7.5 tone/m².

> The critical depth (h) of a channel is determined, $h = \dfrac{v^2}{g}$.

> A dam for storing water is made thicker at the bottom than at the top because pressure on the side increases with depth.

> In any canal, for the uniform flow of water, the canal should be straight at least for 3 miles.

> The bearing capacity of a water logged soils are generally improved by suitable drainage.

> In a hilly region stability of hill slope depends upon angle of slope, ground water condition, geological condition.

> A good quality stone absorbs water less than 5%, most of the stone are having specific gravity ranging between 2.4 to 2.8. Smith's test is performed on stones for soluble and clayey matter.

> *Adit*: Similar to tunnel except that a drift is open to the surface at one end.

> *Raise*: Tunnels made upward for connecting lower level to upper one.

> The area enclosed by ground water divide is described as area of pumping depression.

> For the location of hydroelectric project, survey is conducted in a region during the site investigation is seismic refraction survey.

> The opening that controls the discharge of surplus water from a reservoir is known as *spillway*.

> The magnitude of the uplift pressure in a dam is proportional to the depth of water in the reservoir.

> For the sedimentary beds dipping upstream, the angle which is most competent to support combined load due to weight of dam and pressure of water is 10° to 30°.

> The percolations of water below the dam may be minimised by placing underneath the dam are cut-off walls and sheetpiles.

➢ The process of introducing under suitable pressure, a mixture of cement, sand and water through suitable drill holes in rocks with a view of heating up of cracks and opening j s known as *grouting*.

➢ Tunnel through which water is allowed to flow under a pressure head is the pressure tunnel.

➢ Water which is on the downstream side of the dam is known as *tail water*.

➢ Tunnel constructed for the purpose of water supply is known as *aquiducts*.

➢ In order of increasing specific yield clay-sand-gravel.

➢ *Quants* is a horizontal well, band across the arid region of South Western Asia and South Africa and term synonym to Karez, Foggara or Falaj.

➢ *K* is the coefficient of permeability which depends on the nature of sand and thickness both.

➢ Incase of unconfined aquifer the storage coefficient is equal to specific yield.

➢ 1×10^{-14} is the approximate value of the ionic product of water at 25° (miles/litre).

➢ According to movis and tsui, the maximum capillary rise in inches could be approximated by $hc = \dfrac{2.2}{\alpha \cdot H} \left(\dfrac{\alpha}{1-\alpha} \right)^{1/2}$

➢ If the water is charged with carbon dioxide its solubility in which is 7-fold.

➢ Water table in the upper surface of zone of saturation known as *phreatic surface*.

➢ Suspended water occurs in the zone of aeration and the another name suggested for it is called *vadose water*.

➢ Water which has recently been involved in atmospheric circulation, i.e. it has become a part of hydrogeologic cycle is known as *meteoric water*.

➢ Edme marriotte is regarded as founder of ground water hydrology.

➢ Tunnel contructed for the purpose of water supply is known as *aquiducts*.

➢ Quants os a horizontal well, band across the arid region of South Western Asia and South Africa.

- A saturated formation which field in appreciable quantity of water but through which appreciable leakage of water is possible is called *aquitard*.
- Impermeable formation which neither containnor transmit water is described by the form of aquifuse.
- Meteoric water are typically bicarbonatic waters.
- The water which has remaind out of contact with atmosphere up to present tuine is fossilc water.
- Dating of ground waters is due done by radionucleons present in it.
- A concentrated fossil water or connate water is called *brine*, if the total dissolved solid content is more than 35,000 p.p.m.
- Contiguous zone has a limit of 24 nautical miles from the base line.
- 200 nautical miles is the limit of exclusive ecomic zone.
- Territorial water has a limit of 12 nautical miles.
- Reverse rotatory drill is used for ground water drilling.
- The force of adhesion in capillary water is less than hygroscopic water.
- Water in the intermediate zone when it is nonmoving, is called *pellicular water*.
- The permissible limit of Na (in %) in the ground water for irrigation purpose is 40–60.
- The unit of measurement of radionucleiods present in water is microcurrie.
- Organic matter load in water is determined by BOD value.
- Impermeable formation which may contain water but incapable to transmit it is known as *aquiclude*.
- Capillary water exists as continuous film around the soil particles.
- Sand and gravel in sedimentary deposits often forms perched aquifer.
- Perched aquifer is one which occurs whenever a ground water body is separated from the main ground water by a relatively impermeable stratum of small aerial extent and by the zone of aeration above the main body of ground water.
- A line on water table on each side of which the water table slopes down ward in a direction away from the line is called *devide ground water*.

➤ The difference in levels between the sat (ilatu) water level and the surface of the cone of depression is called *drawdown*.

➤ The piezometric surface of a confined aquifer is, an imaginary surface coinciding with the hydrostatic pressure level in the aquifer, or an imaginary surface parallel to the hydrostatic pressure level of the water in an aquifer or, a unique surface parallel to the hydrostatic level of the water in an aquifer.

➤ Perched aquifer is an unconfined aquifer which occurs within the zone of aeration.

➤ Cappillary water is available to plants.

➤ Piezometric surface may occur below or above the ground water.

➤ An artesion condition of well will occur only when peizometric surface is above the ground surface.

➤ Transmissibility coefficient (T) is equal to coefficient of permeability saturated thickness of aquifer.

➤ When the piezometric surface lie above of ground surface, flowing well result.

➤ Geologic formation and structure that transmit water in sufficient quantity to pumping well or springs is known as *aquifer*.

➤ When a sedimentary formation is subjected to compaction, the porosity decreases with depth.

➤ Porosity which includes only the interconnected pore spaces is known as *effective porosity* and *absolute pore space*.

➤ Percentage of porosity can be given by the relation

$$\frac{\text{Pore volume} \times 100}{\text{Bulk volume}}$$

➤ For drilling ground water wells is extremely hard formation one should use down the hole hammer.

➤ The unit of permeability in CGS system is darcy.

➤ A well graded material is defined as that for which the $\dfrac{d_{60}}{d_{10}}$.

➤ The coefficient of transmissibility is expressed in M²/5 unit.

➤ A uniform material is defined as that for which the $\dfrac{d_{60}}{d_{10}}$

= 2OR < 2.

- The relation between specific yield (Sy), Specific Retention (Sr) and porosity (α), this is shown by $\alpha = Sy + Sr$.
- Water in the zone of aeration is called *vadose water* and *suspended water*.
- Geysers are example of periodic thermal spring.
- If the discharge from the spring is throughout the year, the spring is known as *perinial*.
- An aquifer in which a water table serves as the upper surface of zone of saturation is known as *unconfined aquifer*.
- The moisture content absorbed from an atmosphere of 50% relative. Humidity at 25°C is known as *hygroscopic coefficient*.
- Sonic logging is also known as *porosity logging*.
- For boiler fed, the tolerance limit of total solids in ppm of water at 0–150 PSL pressure is 3000–5000.
- For drinking water, the upper limit of fluoride content is 1.5 ppm.
- The residual moisture content, after the gravitational water is substantially drained out is termed as field capacity.
- A stream recharging ground water of an adjacent unconfined aquifer is known as *influent stream*.
- A stream receiving discharge from ground water is known as *effluent stream*.
- In the Deccan Trap of India, the ground water are mostly contained in weathered zone.
- Storage coefficient in the case of unconfined aquifer, corresponds to its specific field.
- For irritation purpose water is said to be unsuitable when the Na% is > 80.
- Ground water held above an impervious stratum and no connected with main water table is perched water.
- Drilling through the limestone country, if you get a cavern underground from which of the following evidences as revolution per minute of the drilling will be increased.
- In irregular terrain, underground water basin is artificially charged by Ditch or Furrow method.
- In area where the impervious layer is not too far below the ground surface recharging can be done by pits and shafts.
- Confined aquifer are also known *artesian aquifer*.
- Capillary water is held by surface tension.

> Form the total water that falls on the land surface each year, the part which become ground water is about 60%.
> That most suitable method for ground water prospecting is electrical method.
> Underground water, emerging on the surface in forms of seepage and springs, are found to form loose superficial deposits of silica and calcium carbonate near its exists known as *sinters*.
> In drinking water, the iron (Fe) content should not exceed 0.3 ppm.
> Bhabbar is physiographic and hydrologic belt of Ganga–Brahmaputra alluvial province.
> The field of water from individual well in Ganga-Brahmaputra province is 3000–8000 gallons of water per minute.
> Heary water consist of oxygen and heavy hydrogen.
> In soft formation, the method most suitable for ground water prospecting is resistivity method.
> Water saturated unconsolidated sediments shows in crease increase velocity.
> The upper limit of zone of saturation is bounded by water table, and water is known as *unconfined water*.
> At 25°C the total dissolved solids in parts per million of natural water is determined by the formula $\dfrac{TDS}{Ec x 10^6} = 0.64$.
> According to Walton if the d_{50} (effective grain size) of aquifer is more than 0.25 mm and the d_{60}/d_{10} is greater than 3, screened well is used.
> Drinking water should not contain more than 1×10^{-9} microcuries/millilitre strontium and radium are more than 3×10^{-9} μc/millilitre.
> The unit to measure the flow of water is cusec.
> Highest salinity is found in Dead Sea.
> The density of sea water is highest as depth increases and salinity increases.
> The formation of lakes, their pysicochemical conditions and the organism inhabitating them are studied under a special branch of hydrology called *lymnology*.
> Due to the change in barometrie pressure, the level of the rises very slowly at one shore with a corresponding depre-

ssion at the opposite shore in the lake. Such movements are known as *seiches*.

➤ Lysimeter is used to measure transpiration.

➤ Flourides are added to drinking water in order to reduce caries.

➤ A confined aquifer may change to an unconfied aquifer with long pumping, so that the peizometric surface goes below the confining layer (upper).

➤ Sodium absorption ratio (SAR) is mathematically expressed

as $\dfrac{Na^+}{\sqrt{\dfrac{Ca^{++} + Mg^{++}}{2}}}$.

➤ Uniformity coefficient of a soil sample is give by $Cu \dfrac{d_{60}}{d_{10}}$.

➤ Gravel packed wells are used under the circumstances when coefficient is Homogeneous has uniformity coefficient less than 3.0 and an effective gain size less than 0.25 mm.

➤ What percentage of water available on the earth is contained in the ocean 97%.

➤ In the permafrost regions to obtain the underground water, deep wells are required.

➤ Gravel will give the highest specific yields.

➤ Water logging in some particular area may be due to inadequate drainage, over irrigation and seapage from adjoining reservoirs.

➤ The amount of rainfall that enters the ground and penetrates to the ground water table is influenced by geology of the area, hydraulic permeability of ground and turbidity of water.

➤ Steady state flow of incompressible fluids is governed by Bernoullis equation.

➤ If the boundary is a barrier, then the image well is recharging well.

➤ The well loss of a pumping well can be computed with the help step drawdown test.

➤ Due to increase in atmospheric pressure, water table decreases.

➤ As the wind blows across the top of well, water level suddenly rises.

- Due to earthquake, water level remains constant.
- Depression springs are formed where the ground surface intersects the water table.
- Total hardness (TH) is a measure of Ca and Mg content and is customarily expressed as the equivalent of $CaCO_3$. Thus, it is represented by the equation

$$TH = Ca \times \frac{CaCO_3}{Ca} + Mg \times \frac{CaCO_3}{Mg}$$

- The chief source of potassium in ground water is alkali feldspar.
- Subsurface water is more desirable than surface water because the temperature is nearly constant, which is of great importance if the water is used for heat exchange, turbidity and colour are generally absent, chemical composition is commonly constant and radiochemical and biological contamination of most ground water is difficult one.
- The hydrogen ion concentration in pure water 1×10^{-7} moles per litre.
- Conjunctive use of water is a term is hydrology which defines use of both surface and underground water simultaneously for different purposes.
- The term effective porosity refers to the amount of interconnected pore spaces available for fluid flow.
- Water that has been out of hydrologic cycle for at least an appreciable part of a geologic period is termed as connate water.
- Porosity of a formation depends upon shape and size of particles, arrangement of particles and compaction and degree of cementation of individual particles.
- The well made up of silicate rocks can be developed by using hydrofluoric acid.
- Water molecules are held together in crystal by hydrogen bond.
- The maximum diameter of capillary pores are generally 0.05 mm.
- Surveying for artesian resources in any given region involves preparation of water yield maps, permeability maps and hydrochemical maps.

> The most rapid drilling method for unconsolidated formation is reverse circulation rotary method.
> Water yielding capacity is called *storativity*.
> A unit hydrograph usually gives the correlation between direct run-off and the effective rainfall.
> Phosphate coagulation, is used to remove radioactivity of water due to the presence of strontium.
> Flow of fluids are measured by rotatometer.
> Artificial recharge projects are designed to serve to co-ordinate operation of surface and ground water reservoir, provide surface storage for local and important surface water, to reduce or stop significant land subsidence, and to come at adverse conditions such as progressing lowering of ground water level and saline water intrusion.
> Aquiclude is known for clay.
> Aquifuge is known for solid granite.
> Aquitard is known for sandy clay.
> The portion of the rock or soil not occupied by solid mineral matter may be occupied by ground water is called *pore-spaces*.
> Reynold's number is helpful to distinguish between laminar and turbulent flow.
> If V_1 is the velocity of seismic wave above water table and V_2 below it. The depth of water table (H) from ground can be determined by the formula $H = \dfrac{S}{2} \sqrt{\dfrac{V_2 - V_1}{V_2 + V_1}}$.
> For the sedimentary and unconsolidated rocks the value of resistivity lies between $10-10^4$ Ω m^2/m.
> The void ratio of a given soil sample is given by

$$\frac{\text{Volume of voids}}{\text{Volume of soil solids}}.$$

> Zone of aeration consists of interstices occupied partially by water and partially by air.
> Injection of wart water into a deep well can trigger earth-quake.
> The maximum porosity found in sandstone is about 30 per cent.
> The field capacity of any soil in any area depends upon porosity of soil and capillary tension of soil.

- Well log surves as a ready means for obtaining information regarding characters of the rocks or soil formation of the area.
- Due to heavy pumping in confined aquifer, the land surface subsidises.
- A gravel pack in wells stablizes the aquifers, minimizes sand pumping, permits use of a large screen slot with a maximum open area and provides an annular zone of high permeability.
- Renold number (Nr) the Darcy's law is valid for Nr = 1.
- One darcy is equal to 0.987 $(\mu m)^2$.
- Darcy's flow is in the direction of decreasing head.
- According to the Darcy's law, the volume of water passing through the porous media is inversely proportional to the thickness of bed.
- Gravel has highest hydraulic conductivity.
- The sand shape factor "Q" varies from 6.0 to 7.7.
- If pH value of water is seven, water is said to be neutral.
- If pH value of water is less than seven water is said to be acidic.
- If pH value of water is more than 7, water is said to be alkaline.
- When ground water contains 200 mg/litre salts as $CaCO_3$ it is hard.
- Reynold number is the ratio of initial force and viscosity hydraulic radius is equal to area divided by wetted perimeter and hydraulic grade lines for compact clayey soil is 1 and 3 or 4.
- Length of the intruded sea water wedge for a confined aquifer is given by $q = \dfrac{1}{2}\left(\dfrac{Ps - Pf}{Pf}\right)\dfrac{Kb^2}{L}$.
- Ozone is more soluble in water than oxygen.
- Ground water is primary source of water in arid areas.
- The saturated hydraulic conductivity of the soil is given by the formula $\dfrac{Q}{At} = K\dfrac{H}{I}$.
- According to TN Lambe (1951) soils having coefficient of permeability (K) values in the range of 10 to 10 cm/sec has been classified as soils of very low permeability.

➤ Voultation results when turbulent water causes soil particles to nop of skip as they move downward.

➤ Leaching is the process by which fertile soluble part of the soil is dissolved and carried away by water.

➤ The capacity of a soil to retain water is determined by its colour.

➤ Clayey soil retains maximum water.

➤ The smallest volume that can be called soil or smallest three dimensional soil unit which can represent the nature and arrangement of horizon as pedon.

➤ The ability of rock to allow liquids to enter is called *permeability*.

➤ The characteristic of black soil or regur is most fertile soil, it lacks in phosphorous, nitrogen and organic matters, but lime, potash and iron are essential constituent of black soil.

➤ Porosity is greatest in clay.

➤ The soil containing sand, silt and clay in roughly equal proportion is called *loam*.

➤ The clay soil is called cold soil because it dries slowly.

  Resevoirs: Leakage from a reservior is always having practice the trouble, which may be taken place due to faults, fissures, etc. The leakage may be stopped either by grouting or by removing the loose material and refilling the same with cement concrete. The only possible remedies are making provisions for washing out the silt through the passage of the dams, constructing weirs across mouths of the feeding streams and providing a good cover of vegetation on the catchments area.

➤ Indeed the most important thing under the water supply schemes is the selection of source of water, which should be reliable and have minimum number of impurities. After the selection of source of water, the next is to construct intake works to collect it and carry up to treatment plants. At the treatment plant this water will be treated. Types of treatment processes directly depends on the impurities in water at the source and the quality of water required by the consumers. When the water is treated, it is stored in clear water reservoir, from where it will be distributed to the consumers.

➤ The distribution system will also depend on the elevation of clear water reservoir and the elevation of the distribution area. In low level areas water will directly flow under gravitation force, but for high level areas elevated tanks, or pumping will be required.

➤ As the average rainfall in a particular catchments area, rain-guages are fixed at important points. The plan of the whole catchments area is prepared slowing the position of the rain-gauges. The area controlled by each rain-guage is marked on the plan.

➤ All the lines of equal rainfall or precipitation are known as *isohyets* and these lines on the map is called by the name of *isohyetal* map.

➤ By Theissen method, average rainfall over a catchments area is determined. It is in this lines are drawn joining all the neighbouring observation stations. The area of each polygon so formed is multiplied with the average rainfall of that area.

➤ To get the average rainfalls in the whole catchment area, the average of all the rainfalls recorded by various rain-gauges, for which Theissen method is used.

➤ Missing records of one or more stations may be obtained by using the annual figures for the nearest stations multiplied by the ratio of the means of the two stations. For obtaining precipitation during a storm the isolyets of the storm are drawn from the known figures, from which the data which are missing by interpolation method.

Theissen's mean method

$$P = \frac{A_1 P_1 + A_2 P_2 + A_3 P_3 + A_4 P_4 + \cdots + A_n P_n}{A}$$

Where P = mean precipitation on the basin,

A = basin area

$P_1, P_2, P_3, P_4 \ldots\ldots\ldots\ldots P_n$ = precipitation at the respective stations whose surrounding polygons have the areas $A_1, A_2, A_3, A_4 \ldots\ldots\ldots A_n$ respectively.

➤ As the precipitation over an area varies from day-to-day, month-to-month and even from year-to-year, are recorded accurately by the Meteorological departments.

➤ The maximum intensity of rainfall will help in determining the maximum quantity of water, which will flow in the rivers, lakes, etc. The design of collection works, bridges, culverts and other structures will directly depends on the maximum quantity of run-off.

➤ The duration of the dry weather and the availability of the water in that period will help in deciding the actual size of the storage reservoir. For this also at least dry weather records of three successive dry years should be taken into account.

➤ The rate of precipitation over a particular area, and its duration are very important for several reasons. The intensity of rainfall in centimetres per hour as :

$$i = \frac{T+1}{t+1} \times \frac{F}{T}$$

Where i = Intensity of rainfall in centimetres.

      T = Duration of the precipitation in hours.

      t = The time in hours for which the intensity of rainfall remains maximum.

      F = Total rainfall in centimetres.

As from the above formula, if t will be shorter, the intensity of rainfall will be higher. But it t is short, the time taken for run-off from the distant part of the catchment area will be more and it will not reach at the site of the bridge.

➤ The intensity of precipitation for which the discharge is maximum when it remains for a concentrated period it is called by the name as the critical intensity of the rainfall.

➤ The greater is the intensity of precipitation shorter is the duration. In fact, it is not necessary that high intensity precipitation will give maximum discharge. In practice it has also been observed that a high intensity storm will have lower frequency, i.e. it will be repeated again after long duration of time interval sufficient local data is available.

$$R = \frac{N}{M - 0.5}$$

Where, R = Recurrence interval or the average period in years.

      N = Length of record in years.

      M = Number of each event or intensity when arranged in descending order of magnitude.

> (a) The run-off mainly depends on the area of the catchment.
> (b) Slope and shape of the catchments area.
> (c) The degree of porosity of the soil of the catchment area.
> (d) Obstruction in the flow of water due to trees fields and gardens, etc.
> (e) Initial state of the catchment area with respect to the wetness.

> For determining evaporation loss : $\Rightarrow$
> Rohwer's formula:
>
> $E = c' (1.465 - 0.00732p) (0.44 + 0.0732u) (v_1 - v_2)$.
>
> where, $c = 0.750$ approx. and is determined from the available evaporation data.
>
> P = Atmospheric pressure (in cm mercury) at a 0°C temperature.
>
> Eiv, v_1 and v_2 have the same measurement.

> Maharashtra Engineering Research Institute at Bhatghar and Khadakwasla and Pashan, after research have the following formula for the calculation of evaporation losses.
>
> $E_i = 0.4 + 0.04.u + 0.50 (e_a - e_d)$ cm/day.
>
> e_a = Mean vapour pressure (in cm of mercury) of saturated air at water surface temperature.
>
> e_d = Mean vapour pressure (in cm of mercury) at dew point temperature.
>
> E_i and v have the same meaning as the above formula. It is also considered as :
>
> $$E_i = \frac{76600}{T^{3.369}} (e_a - e_d) \text{ cm/day}$$
>
> Where, T = water surface temperature in degree centigrade.
> E_i, e_a and e_d have the some meanings as above.

> For the run-off determination, it is in the form of streams or rivers. The total quantity of run-off which reaches at a particular point of the stream can be determined directly by measuring the discharge of the stream.

> For an economic and efficient design of flood control measures floods are to be estimated with reasonable accuracy. The ideal solution would be achieved in all the above cases if the hydrologist could supply the hydraulic engineer with an estimate of maximum flood which would occur at the site under consideration during a specific period of time.

Bibliography

1. Anon (1971): Geophysics and ground water. *Water Well Jour.*, **25**, No. 07, pp. 43-60, No. 08, pp. 35-50, 1971.

2. Ackermann, N.L. and Chang Y.Y. (1971): Salt water interface during ground water pumping. *Jour. Hydraulics Div., Amer. Soc.Civil Engrs*, **97**, No. Hy2, pp. 223-232.

3. Archyana (1975): Transform fault and rift valley from bathyscaph and diving saucer. *Science*, **190**, pp. 108–16.

4. Biemond, C. (1957): Dune water flow and replenishment in the catchment area of the amsterdam water supply. *Jour, Instn, Water Engrs*, **11**, pp. 195-213.

5. Bennett, G.D. and Pattern E.P., Jr.(1960): Borehole geophysical methods for analysing specific capacity of multiquifer wells. *U.S. Geological Survey Water-Supply Paper* 1536 A, p. 25.

6. Bear J. and Dagan G. (1964): Intercepting fresh-water above the interface in a coastal aquifer. *Intl. Assoc. Sci. Hydrology Publ.* 64, pp. 154-181.

7. Banatti, E (1967): Mechanisms of deep-sea volcanism in the South Pacific, *Researches in Geochemistry*, Vol. 2, John Wiley & Sons, Inc., New York, pp. 453-91.

8. Blankennagel, R.K. (1968): Geophysical logging and hydraulic testing, Pahute Mesa, Nevada Test Site, *Ground Water,* **6**, No. 4, pp. 24-31.

9. Bruington, A.E. (1969): Control of salt-water intrusion in a ground water aquifer, *Ground Water*, **7**, No. 03, pp. 9-14.

10. Brown, R.H. et al (1972): Ground water studies: Studies and Reports. In *Hydrology* **7**, UNESCO, Paris.

11. Bowden, L.W. and Pruits E.L. (Eds) (1975): *Manual of Remote Sensing,* Vol. II, *Interpretation and Applications*, Amer Soc. Photogrametry, Falls Church, Virginia, pp. 869-2144.

12. Bugg, S.F. and Lolyd J.W. (1976): A study of fresh-water lens configuration on the Cayman islands using resistively methods, *Quarterly Jour. Engring Geol.*, **9**, pp. 291-302.

13. Ballare, R.D., Bryan, W.B., Heirtzler, J.R., Keller, G., Moore, J.G. and Van Andel, T. (1975): Manned submersibe observations in the *famous* area, mid-Atlantic ridge, *Science*, **190**, pp. 103-8

14. Ballard, R.D and Van Andel, T.H. (1977): Morophology and tectonics of the inner rift valley at lat 36°50' N on the mid-Atlantic ridge: *Geol. Soc. America Bull*, **88**, pp. 507-30.

15. California Dept. (1958): Water Resources, Santa Ana Gap salinity barrier. *Orange Country Bull,* 147-1, Sacramento, pp.178.

16. Compagnie Generale de Geophysique (1963): *Master Curves for Electrical Sounding,* 2nd edn, European Assoc. Exploration Geophysicists, The Hague, Netherlands, 49 pp.

17. Cahill, J.M. (1967): Hydraulic sand-model study of the cyclic flow of salt water in a coastal aquifer, U.S. Geological Survey Prof. Paper 575-B, pp. 240-244.

18. Curray, J.R. (Ed.) (1969): New concepts of continental margin sedimentation, AGI short course notes, Amer. Geol. Inst., Washington, DC, pp. 340.

19. Cartwright, K. : (1970): Ground water discharge in the Illinois Basin as suggested by temperature anomalies. *Water Resources Research,* **6**, No. 03, pp. 912-918.

20. Chadwick, D.G. and Jensen L. (1971): The detection of magnetic fields caused by ground water and the correction of such fields with water dowsing, Utah Water Research Lab, Logan, Jan. 1971, pp.57.

21. Crosby, J.W. and Anderson J.V. (1971): Some applications of geophysical well logging to basalt hydrogeology, *Ground Water,* **9**, No. 05, pp.12 20.

22. Chandler, R.A. and McWhorter D.B. (1975): Upconing of the salt-water fresh-water interface beneath a pumping well, *Ground Water*, **13**, pp. 354-359.

23. Chidley, T.R.E. and Lolyd J.W. (1977): A mathematical model study of fresh-water lenses, *Ground Water*, **15**, pp. 215-222.

24. Dietz, R.S. (1964): Origin of continental slopes: *Amer. Scientist*, **52**, pp. 50-69.

25. Dobrin, M.B. (1976): *Introduction to Geophysical Prospecting*, 3rd edn, McGraw-Hill, New York, pp. 630.

26. Emery, K.O. (1968): Relict sediments on continental shelves of the world. *Amer. Assoc. Petroleum Geologists Bull.*, **52**, pp. 445-64.

27. Emery, K.O. (1969): The continental shelves. *Sci. American*, **221**, No. 3, pp. 106-22.

28. Ellis A.J. (1917): The divining rod—a history of water witching. U.S. Geological Survey Water-Supply Paper 416, pp. 59.

29. Emery, K.O., Uchupi, E., Phillips, J.D. Bowin, C.O., Bunce, E.T. and Knott S.T. (1970) : Continental rise off eastern North America. *Amer. Assoc. Petroleum Geologists Bull.*, **54**, pp. 44-108.

30. Emery, K.O. and Gunnerson, C.G. (1973): Internal Swash and surf. *NatlAcad. Sci. Proc.*, 70, pp. 2379–80.

31. Flathe, H. (1963): Five-layer master curves for the hydrogeological interpretation of geoelecrtical receptivity measurement above a two-storey aquifer. *Geophys. Prospecting*, **11**, pp. 471-508.

32. Frimpter, M.H. (1969): Casing detector and self potential logger, Ground Water, **07**, No. 6, pp. 24-27.

33. Gorder, Z.A. (1963): Television inspection of a gravel pack well, *Jour. Amer. Water Works Assoc.*, **55**, pp. 31–34.

34. Griffiths, D.H. and King R.F. (1965): *Applied Geophysics for Engineers and Geologists*, Pergamon, Oxford, pp. 223.

35. Hess, H.H. (1946): Drowned ancient islands of the pacific basin. *Amer. Jour Sci.*, **224**, pp. 772-91.

36. Heezen, B.C. and Ewing, M. (1952): Turbidity currents and

submarine slumps and the 1929 Grand Banks earthquake. *Amer. Jour. Sci,* **250**, pp. 849-73.

37. Hamilton, E.L. (1956): Sunken Island of the Mid-Pacific Mountains. *Geol Soc. America Mem.* 64, pp. 97.

38. HSU, K.J. (1972): When the mediterranean dried up. *Sci. American*, **227**, No. 06, pp. 26-36.

39. Hollister, C.D. (1973): Atlantic continental shelf and slope of the united states-texture of surface sediments from New Jersey to Southern Florida: U.S. Geol. Survey Prof. Paper 529-M, pp. 23.

40. Jannasch, H.W., Eimhjellen, K.C.O. and Farman Farmaian, A. (1971) : Microbial degradation of organic matter in the deep sea. *Science*, **171**, pp. 672-75.

41. Karig, D.E. (1974): Evolution are systems in the western pacific. *Ann. Rev. Earth and Planetary Sci.*, **2**, pp. 51-75.

42. Luyendyk, B.P. (1970): Origin and history of abyssal hills in the north-earth pacific ocean. *Geol. Soc. America Bull.*, **81**, pp. 2237-60.

Index

GEOMORPHOLOGY
AND
HYDROGEOLOGY

A Handa